MODERN APPROACHES IN FOREST ECOSYSTEM MODELLING

EUROPEAN FOREST INSTITUTE RESEARCH REPORT

NO. 8

MODERN APPROACHES IN FOREST ECOSYSTEM MODELLING

European Forest Institute Research Report 8

BY

OLEG G. CHERTOV
ALEXANDER S. KOMAROV
GEORGY P. KAREV

BRILL
LEIDEN · BOSTON · KÖLN
1999

The views expressed in this book are those of the authors and do not necessarily correspond to those of the European Forest Institute.

This book is printed on acid-free paper.

Library of Congress Cataloging-in-Publication Data

Modern approaches in forest ecosystem modelling / by Oleg G. Chertov, Alexander S. Komarov, Georgy P. Karev.
p. cm. – (European Forest Institute research report, ISSN 1238-8785 ; 8)
Includes bibliographical references and index.
ISBN 9004114157 (cloth : alk. paper)
1. Forest ecology—Simulation methods. I. Chertov, O. G. (Oleg Georgievich) II. Komarov, A. S. (Aleksandr Sergeevich) III. Karev, Georgy P. IV. European Forest Institute. V. Series: Research report (European Forest Institute) ; 8.
QK938.F6M585 1999
577.3'01'1—dc21 99–12155
CIP

Die Deutsche Bibliothek - CIP-Einheitsaufnahme

Modern approaches in forest ecosystem modelling / by Oleg G. Chertov ... – Leiden ; Boston ; Köln : Brill, 1999
(European Forest Institute research report ; No. 8)
ISBN 90–04–11415–7

ISSN 1238-8785
ISBN 90 04 11415 7

PRINTED IN THE NETHERLANDS

Contents

Preface

The idea of this review originated during work on forest ecosystem modelling at the European Forest Institute by the authors. It was one of the additional results of the exciting work on the creation of a forest ecosystem model for European boreal forests (Chertov and Komarov, 1995b; 1997b).

The review has been written by a team of experts in various sciences. The section on simulation models was written by a forest ecologist, Dr. O. Chertov; and that on spatial models by an ecological modeller and mathematician, Dr. A. Komarov. The review of analytical models and approaches was prepared by a mathematician and biological modeller, Dr. G. Karev. Section 1.3.7 in Chapter 1 was written by A. Komarov and Section 3.5 in Chapter 3 by O. Chertov.

It is necessary to emphasize that the sections of the review represent different points of view about the problems of forest modelling. We think that this is a positive progressive feature of the review, reflecting the great diversity of theoretical and methodological approaches in this very intensively developing field. Moreover, some items are duplicated (gap, individual-based and regional models) in the sections on simulation and analytical modelling. These items are of great importance and they occur throughout the review, with discussion of the problems from an ecological and a mathematical point of view. Every section of the review contains a description of some of the authors' ideas and results, reflecting their personal theoretical and methodological positions in forest modelling.

The review reflects the state-of-the-art as a whole, with an emphasis on some Russian approaches that are poorly known in the West. Three decades ago there were a large number of young professional mathematicians in the new field of ecological modelling in Russia. The development of modelling was therefore strongly influenced by mathematics and physics, and as a result, there has been vigorous development of theoretical approaches in forest modelling. Thus the structure of this review reflects the history of forest modelling in Russia, with special attention being given to theoretical approaches. We should note that Russian works cited are mostly published by Russian Academy of Sciences (academic journals and Nauka Publishers) and are available. The review is designed for two groups of readers: *(i)* forest ecologists and modellers (Chapter 1); *(ii)* theoretical ecologists and applied mathematicians (Chapters 2 and 3).

This is an analytical review and not a register of all models discussed, with their technical description. Registers of models are currently on the Internet and include quite comprehensive information of this type. We can mention:

- Register of Ecological Models by the University of Kassel (http://dino.wiz.uni-kassel.de/model_db/server.html);
- CAMASE Register of AgroEcological Models (Plentiger and Penning de Vries, 1996) and Database at WWW (http://www.bib.wau.nl/cgi-bin/mini.sis/camasall?A010=@);
- GSTE SOMNET Report of soil organic matter models (Smith et al., 1996) and Database at WWW (http://yacorba.res.bbsrc.ac.uk/cgi.bin/somnet).

The authors hope that the particular features of the review will allow the materials and opinions discussed to be useful for further development of forest modelling, both as a practical tool for environmental forest management in the next century and as a theoretical tool in forest science and ecology.

Acknowledgments

The work has been supported by European Forest Institute Project 605. The authors express their great thanks to the EFI Secretariat and all the staff for the exciting and friendly atmosphere at the Institute during the preparation of the review and to Dr. P. Grabarnik for valuable comments to Chapter 2. Special gratitude should be expressed to the manuscript reviewers, Prof. G.I. Ågren, Prof. M.G.R. Cannell and Dr. M. Lindner, and to Dr. M. Jarvis who revised and edited the language.

Introduction

Forest modelling is currently developing intensively in forest ecology. Among much new literature containing original descriptions and applications of models, there are many recent reviews, comparative studies, registers of models and special issues in international journals (e.g. Dale et al., 1985; Ågren et al., 1991; Antonovsky et al., 1991; Mohren and Kienast, 1991; Berezovskaya and Karev, 1994; Breckling and Müller, 1994; Liu and Ashton, 1995; Tiktak and Van Grinsven, 1995; Päivinen and Nabuurs, 1996). There have been extensive discussions of trends and prospects in forest ecosystem modelling (Huston et al., 1988; Shugart et al., 1992a; Levine et al., 1993). These circumstances make the writing of a new review on the subject very complex. Nevertheless we decided to implement the idea because we have our own experience and views of the problems, which are different from those previously published. In addition, we perceived a need to mention some Russian ideas, concepts and approaches in the field, which are not well known in the West. An additional aim of the review is, therefore, to explain the Russian state-of-the-art in forest modelling. There have been some comprehensive reviews published in Russian on forest modelling in recent decades (Terskov and Terskova, 1980; Oja, 1985a; Kull and Kull, 1989; Berezovskaya et al., 1991; Korzukhin and Semevsky, 1992).

The current state of mathematical modelling of forest ecosystems is determined by three different scientific approaches. The layer-mosaic concept (gap paradigm) of the spatial-age structure of forest ecosystems was generated in theoretical ecology. According to the concept a driving force of forest development is a formation of the gaps of regeneration in forest canopy on sites of fallen trees. A broad class of simulation gap-models has been constructed with the advent of computer modelling of forest ecosystems. The theory of structured models of populations, linking behaviour of a population as a whole with the dynamics and interaction of individuals, is well advanced in mathematical biology and analytical models of forest populations and communities have been developed and studied on the basis of this theory.

It seems to us that there are the following growth points in forest simulation modelling, reflected in the review below: the individual-based approach; the modelling of spatial structures; combination (combined, hybrid) modelling, especially for a tree-soil system; and regional (landscape) modelling.

The individual-based approach is currently an intensively developing approach in the modelling of various ecological systems (Metz and Deikman, 1986; Levin et al., 1989; DeAngelis and Gross, 1991). Individual-based or structural models are adequate tools for linking behaviour of the population as a whole with dynamics of individuals and their interactions. The models represent the dynamics of the ensemble of trajectories of the individuals. The analytical models analyse the main properties of the probabilistic distributions of parameters and the evolution and asymptotic behaviour of the initial distribution of individuals; the simulation models calculate directly the fate of each individual or the dynamics of some integrated unit, such as an age cohort. Gap-modelling of forest ecosystems is one of the main fields of application of both simulation (Botkin, 1993; Shugart, 1984, Shugart et al., 1992a) and analytical (Berezovskaya and Karev, 1990, Karev, 1994) individual-based models.

Spatial structure modelling has largely been derived from the individual-based ideology. Trees and plants are sessile organisms. Thus interactions between individuals that influence the growth parameters of a tree depend on its position within a stand. Density effects are well-known consequences of these nearest-neighbour interactions leading to bimodality of distribution of a tree's characteristics (Ford, 1975; Huston and DeAngelis, 1987). Interactions may be a result of shading and/or redistribution of the soil and water resources among the trees. Some concepts of constructive quantitative descriptions of these interactions, such as the "phytogenic field" concept (Uranov, 1965) or "ecological field theory" (Wu et al., 1985), have been suggested. Spatial problems are mostly studied by simulation modelling, with the exception of some work with probabilistic models (Diggle, 1976; Gates, 1978). In the last two decades, a new tendency in analysing the spatial structure of a forest has developed, concerned with methods of spatial statistics (Komarov, 1979; Grabarnik and Komarov, 1981; Tomppo and Sarkka, 1995).

Combination (combined, hybrid) modelling is a recently formulated approach, which is a further evolution of the individual-based approach (Huston et al., 1988; Post and Pastor, 1990; Levine et al., 1993). This involves a combination of the initial structural model and processes models. An important example is the "forest stand – soil" system, taking into consideration the role of ground vegetation and, perhaps, the understorey. Combination models should be an instrument for biological cycle simulation keeping an account of the balances of carbon, nitrogen and other elements and taking into consideration the interactions between individual trees and other plants with respect to redistribution of restricted, continuously distributed resources for growth (Chertov and Komarov, 1995b, 1997b).

Regional (landscape) modelling seems not to be such an intensively developed field of forest simulation as are stand models of various types. However, these models are indispensable for forestry and environmental planning and management. There are currently some successful attempts at regional modelling (Urban et al., 1991; Shugart et al., 1992a; Acevedo et al., 1995). One of the main difficulties of *landscape modelling* is enlargement of the spatial-temporal scales of the models. Our opinion is that this task may be solved only through the construction of a hierarchical system of forest models corresponding to different spatial and temporal scales and various levels of structural organisation.

Detailed simulation models are characterised by a large number of variables and parameters, as well as by a number of coefficients, to describe a definite object (i.e. a tree of a certain species, a forest in a given geographical and climatic zone) as precisely as possible, together with forecasting growth and development over rather short intervals of time. The size and complexity of these models does not usually allow analysis of the influence of a large number of parameters on the qualitative behaviour of the model (i.e. sensitivity analysis). To carry out this task, it is convenient to apply analytical models to the description of the dynamics of a few essential variables and parameters and their qualitative peculiarities, as well as to study of limit regimes and changes in the spatial boundaries. It seems promising to apply jointly both analytical and simulation approaches, the analytical models being regarded as sub routines of the simulation models. This makes it possible to adapt models to the modelled objects, to forecast likely situations and, finally, to assess the precision of the simulation approach. However the benefit and necessity of analytical models is far from being exhausted and further developments are urgently needed for particular processes.

In this review we want to show that the results of these approaches may be considered as components of a united theory, which is based on, and was generated by, the mathematical theory of structured models. We hope that this new theory will allow investigation of the dynamics of the forest ecosystem on various spatial-temporal scales. The main goal of this theory is to develop methods of construction and investigation of a hierarchical systems of models based on different approaches and at various spatial and temporal scales (following the principles of system analysis), where models of higher levels include models of lower levels as sub models or elementary objects. A hierarchical system of forest population and ecosystem models comprises modules of increasing structural, spatial and temporal complexity: tree – gap (locus) – population – metapopulation – forest ecosystem – forest territory (landscape, region) and then, possibly, – geographical zone (ecoregion) – continent (subcontinent) – globe. The aim of such a hierarchical scheme is not to include each and every detail, but to link models across scales, so as to be able to calibrate more aggregate models using detailed analysis at a smaller scale, and also to obtain consistency in the analysis of ecosystem dynamics in a hierarchical approach.

In the chapter devoted to simulation models (Chapter 1), there is special reference to soil organic matter (SOM) models. The SOM dynamics is one of the main macro-processes in the edaphic environment being a) a destructive component in forest ecosystems responsible for growth rate linked to the cycle of nutrients and water, and b) a stabilising component resulting from SOM accumulation. Moreover, the development of a SOM sub model is a driving factor underlying ecogenetical (primary) succession, although this is not really of interest to the majority of forest modellers. The theoretical outputs provide the main stimulation for a discussion of analytical models as being important for simulation modelling. In this review, we do not consider the empirical regression models previously developed and popular in forest ecology, and still widely used in forest mensuration. Nor do we consider large ecological-economic model systems. They are, perhaps, a matter for the immediate future, and must be a synthesis of ecological or silvicultural models of different scales with economic models. Existing large-scale scenario models (Päivinen and Nabuurs, 1996) can be a prototype for such an approach.

The structure of this review follows the general classification of the models as "simulation" or "analytical", with intermediate types as "spatial" models, and considers all the topics mentioned above. Simulation models use numerical solutions on computers, analytical models operate with methods of classical mathematical analysis. However, it is sometimes very difficult to assign a model to one of these groups. Our classification divides all the models on these two groups according to the relative proportion of numerical and mathematical methods used.

Finally we should point out that the main idea of the review is to state a transition of forest modelling from an art to the precise science which is clearly seen nowadays. From a methodological point of view there is now a shift from theoretical analytical modelling of continuous variables to simulation modelling of a set of discrete objects with local interactions, including spatially explicit models. Moreover, we believe that forest ecosystem modelling will be a new practical tool in environmentally oriented, sustainable forestry of the near future, allowing precise, realistic and comprehensive prediction of the growth of forest stands and environmental dynamics. We expect that the forest ecosystem models will replace traditional growth tables in 3rd millennium.

1 Simulation Models of Forest Ecosystems

1.1 Overview

At the present time there are a large number of different models considering both the whole ecosystem and the components of various ecosystems or stands. For example, there are models of tree crown structure (architectonics), and there are models of forest watersheds. We have concentrated our efforts on description of the main model approaches on the spatial scales from a single tree to a stand (forest ecosystem) and landscape. Thus we do not go into detailed descriptions of ecophysiological models of photosynthesis, on the one hand, or continental or global models of terrestrial ecosystems, on the other. Neither is there any discussion of forest hydrological and geochemical models. However, we give special attention to soil organic matter models that deal with the crucial role of forest soil as both a destructive and a stabilising component in forest ecosystem functioning.

1.2 Models of Soil Organic Matter Dynamics

1.2.1 Introduction

From an ecological point of view any soil represents a destructive component of terrestrial and semi-aquatic ecosystems. The process of photosynthesis binds solar energy and chemical elements in plant biomass and the soil plays an opposite role. The soil is responsible for decomposition of synthesised organic matter with release of energy and elements. Released energy is dissipated, but released elements and compounds (CO_2, N, P, K, S, Ca, Mg and many others) are mostly re-utilised by organisms in the ecosystem, in a biological cycle. Therefore, the rates of destructive processes in a soil are factors limiting the rate of photosynthesis.

The other important function of the soil is a constructive role, such as the accumulation in the top soil of humified organic matter. This is an important process that increases an ecosystem's stability, as a result of the presence of a sufficient pool of nutrient elements accumulated in the soil organic matter (SOM).

The necessity to model the soil system had already been expressed in the last century by V.V. Dokuchaev and repeated by Jenny et al. (1949). However, the first attempts at analysis of soil functioning and discussion of the prospects for mathematical modelling of the soil system were published in the 1970s (Bondarenko and Liapunov, 1973; Kline, 1973; Runge, 1973, Hugget, 1975; Yaalon, 1975). Some work with different methodological approaches to soil mathematical simulation appeared at that time (Bazilevich, 1978; Chertov et al., 1978; Ulrich et al., 1979). The problem is also currently being discussed (Smeck et al., 1983; Lavelle, 1987; Sverdrup, 1990; Addiscott, 1993; Bouma and Hak-Ten Broecke, 1993; Chertov and Komarov, 1995a).

The process of soil formation is presently considered as the consequences of the impact of solid-phase ecosystem debris on the surface layers of parent rocks (Targulian, 1986, 1987). In agreement with this concept, soil can be represented as a two-compartment system as a first approximation: a sub-system of soil organic matter transformation and a sub-system of parent material transformation, i.e. weathering. Transformation of mineral substrata (parent rock weathering) is essential for release of the majority of nutrient elements (with the exception of nitrogen) and the formation of loose parent material rich in clay minerals.

From the point of view of ecosystem functioning, the SOM subsystem is dominant in forests for three reasons: (i) SOM transformation is a main macro-process in topsoil, determining practically all the other soil processes and properties; (ii) nitrogen is the principal limiting variable in the edaphic environment of forest ecosystems; its kinetics are closely connected with SOM dynamics and determine plant growth; (iii) the rate of SOM transformation corresponds to the rate of stand dynamics, being coincident in time; at the same time the rate of mineral phase weathering in a soil is slow, and released elements (P, Ca, Mg, S etc.) are finally also accumulated in SOM.

The quantification of SOM dynamics is of great significance from both ecological and pedological points of view. The first efforts had already been made in the 19th century (Kostychev, 1889). Then they were repeated in the 1930s (Tiurin, 1937). The Jenny et al. (1949) negative exponential function is the classic, well-known example of the earliest SOM model.

1.2.2 Early SOM models

The most widely used model in forest ecology to date is the Olson (1963) model. This model considers decomposition of debris as a process of mineralization only, and does not include a description of humification. Prusinkiewicz (1977) made the first attempt to model forest SOM on the basis of the classic concept of humus types (mor, moder, mull). His model accentuates energy flows during litter transformation, accomplished by three main groups of soil microorganisms, and formally represented as three components of humus with different rates of decomposition. The model of Nakane (1978), intended for description of SOM dynamics in natural ecosystems, takes into consideration litter mineralization, humification and, additionally, the vertical distribution of humified material.

SOM simulation in grassland ecosystems began at the same time but the models differed with respect to the main processes and areas of application. The simple model of Gilmanov (1974) represents quite an interesting approach based on comprehensive data on biological productivity. A more detailed SOM model was elaborated by Hunt (1977). His model included ten components of surface and belowground litter inputs, three components of humic materials at different soil depths, and eight components of microbial decomposers.

There were also some successful efforts in agricultural SOM modelling in that period, with regard to SOM transformation in mineral top soil resulting from inputs of both crop residues and organic manure (Jenkinson and Rayner, 1977; Gonchar-Zaikin and Zhuravlev, 1979; Smith, 1979).

It is important to point out that all SOM models of the 1960s and 1970s embodied application of microbiological theory of SOM transformation, including humification. The only exception was a model of Prusinkiewicz (1977), who highlighted the role of soil fauna. At the same time, the SOM modelling originated largely from the development of ecosystem models (and not from the development of pedological theory).

1.2.3 Recent SOM models

Nowadays there are a number of SOM models reflecting different theoretical approaches and practical uses. Models of soil nitrogen dynamics developed simultaneously, but we do not discuss them here.

We have listed the main promising SOM models to extend understanding of their structure and possibilities (Table 1).

Most of the models discussed below consider SOM dynamics as dependent on quality and quantity of litter (and organic manure), temperature, and on parameters reflecting the water regime. Some models require direct input data on soil moisture (Li et al., 1994; Franko et al., 1995; Chertov and Komarov, 1996); others need data on evapotranspiration, being a module of ecosystem models, e.g. the well-known LINKAGE (Pastor and Post, 1985) and CENTURY (Parton et al., 1988) models. There is also a group of models in which soil texture has an influence (Parton et al., 1988; Jenkinson, 1990; Verberne et al., 1990; Hansen et al., 1991; Li et al., 1994; Franko et al., 1995).

The forest productivity-soil process model LINKAGE (Pastor and Post, 1985) is a well-argued, ecosystem-oriented SOM model. The model simulates the mineralization of SOM as a one-pool system with special reference to available nitrogen supply for tree growth. A particular feature of the model is the use of actual evapotranspiration as an integrated parameter determining the rate of SOM mineralization. This model is now widely used. For example, it has been used by Kellomäki et al. (1992) as a component in a boreal forest model that combines a process-based model with a gap-type one.

The modelling approach of Ågren and Bosatta (1987; Bosatta and Ågren, 1991) is based on a comprehensive study of forest litter decomposition in Sweden and North semisphere. It takes into account variation in the fine details of litter composition and changes in the course of SOM transformation and some general ideas of substrate and decomposer interactions. The model has been developed primarily for forest ecosystems.

Rather different concepts have been realised in the forest plantation model (ITE model) of Thornley and Cannell (1992). This is quite a simple approach with representation of SOM in one pool only, plus a soil organisms compartment. The model has four components of litter fall and three of mineral nitrogen. It treats mineralization only, with detailed presentation of nitrogen transformation (ammonification and nitrification), but humification is not taken into consideration. The model has been elaborated primarily as a theoretical tool, but it is now being used for simulation of SOM dynamics and tree nutrition.

The model of Goto et al. (1994) is an extension of the Nakane

Table 1. Characteristics of some recent SOM models.

Models	Processes considered				Components		Applications		
	Mineralization by soil micro-organisms	Mineralization by soil micro-organisms and soil fauna	Mineralization and humification by soil micro-organisms	Mineralization and humification by micro-organisms and soil fauna	SOM fractions only	Fractions of SOM and soil biota	Theoretical: mostly carbon balance	Agro-ecosystems and grasslands	Forest ecosystems
Molina et al., 1983			X			X		X	
Pastor and Post, 1985	X				X		X		X
Ågren and Bosatta, 1987; 1996	X					X	X		X
Hunt et al., 1987		X				X			
Moran et al., 1988		X				X	X	X	
Parton et al., 1988			X			X	X	X	
Jenkinson, 1990			X			X	X	X	
Verberne et al., 1990			X			X		X	
Hansen et al., 1991			X			X		X	
Thornley and Cannell, 1992	X					X	X		X
Grant et al., 1993a,b	X					X	X	X	
Ryzhova, 1993a,b			X		X		X		
Goto et al., 1994			X		X		X		
Li et al., 1994			X			X		X	
Franko et al., 1995			X		X			X	
Chertov and Komarov, 1996, 1997				X	X		X		X

model. SOM occurs in the model as dead biomass and humus. The model has two components of litter fall, and embraces humification of dead biomass, decomposition of dead biomass and humus, and, finally, downward movement of humus in the soil profile. This is an universal model allowing simulation of SOM dynamics in a wide range of terrestrial ecosystems. There are other SOM models of natural ecosystems based on similar approaches, e.g. the model NAM-SOM (Ryzhova, 1993a,b).

The CENTURY model (Parton et al., 1988; Paustian et al., 1992) is an SOM model of grassland ecosystems of the greatest interest. CENTURY has two components of litter input (structural and metabolic), with different rates of decomposition, and three of SOM carbon (active, including biomass of micro-organisms, slow, and passive). This model considers dynamics of carbon, nitrogen and other elements and has been used in simulation of grassland ecosystems and now also for predictions of effects of climate change on the global SOM pool (Potter et al., 1993). The model has been comprehensively validated for agricultural soils, too.

At the present time, the assembly of SOM models is going on very intensively in agronomic pedology. The well-known ROTHC model (Coleman and Jenkinson, 1995) is based on classical, long-term agricultural experiments at Rothamsted Experimental Station (Jenkinson et al., 1987; Jenkinson, 1990). The model has two inputs of litter and a cascade of repeated pairs of biomass (microorganisms) and humified material, distinguished by the rate of SOM mineralization. The additional specific feature of ROTHC is a pool of inert organic matter (IOM) that is not a part of the SOM turnover. This pool is included in some other models, for example in DAISY (Hansen et al., 1991), taking into consideration litter and SOM C/N ratios, and transformation of nitrogen with more detailed SOM compartmentalization. The CANDY model (Franko et al., 1995) is similar in some respects (for example, a fraction of inert organic matter), as are the models NCSOIL (Molina et al., 1983; Nikolardot et al., 1994) and VVV (Table 1, van Veen et al., 1984; Verberne et al., 1990), which are of the same kind.

The very detailed model of Grant et al. (1993a,b) contains a detailed presentation of the functional organization of four litter and SOM pools comprising the mass of SOM, biomass of microorganisms, soluble organic matter and microbial residues, and represents the processes of SOM and biomass transformations in every SOM pool. This model is very complicated, allowing calculation of the dynamics of SOM carbon and nitrogen, i.e. mineralization, immobilization and retention, intermediate products of decomposition, and microbial biomass. The model has been verified by laboratory experiments in controlled conditions. The DNDC model (Li et al., 1994) is very similar to that of Grant et al. and considers activity of different groups of microorganisms in the process of SOM mineralization. This model is a component of a full agro-ecosystem model.

The authors' current model, SOMM (Chertov and Komarov, 1995a, 1996, 1997a), based on earlier simulation of forest floor mineralization and humification (Chertov, 1985), has been developed using the classic concept of humus types. The model has one pool of litter fall and three of SOM: undecomposed litter, partly humified organic material (forest floor and peat), and humus bonded with the mineral matrix of the top soil. The litter input in the model can be represented by other fractions, for example by dead wood, which have characteristic, species-specific, ash and nitrogen contents. There are three processes of SOM humification by three communities of micro-organisms (without consideration of their biomass) and three processes of mineralization. The model is represented by a system of ordinary differential equations with coefficients that depend on soil temperature and moisture, litter nitrogen and ash content, and on the C/N ratio in the mineral topsoil. The dependencies of the coefficients on the above variables are taken from experimental data on litter decomposition in various conditions. The model also considers nitrogen transformation and release.

The main features of the SOMM model are: *(i)* correspondence with the classical concepts of humus types; *(ii)* the minimal set of input parameters (quantity of litter; species-specific characterisation of litter quality: nitrogen and ash contents, soil temperature and moisture); and *(iii)* involvement of the soil fauna as a component of consequence, and their interactions with micro-organisms in the biochemical processes of litter decomposition. It is noteworthy that the main inter-dependencies between the decomposition activities of the soil fauna and climatic conditions in SOMM are the same for different geographic zones (Chertov and Komarov, 1997a). SOMM has been used for wide-scale simulation of SOM dynamics in forest ecosystem modelling and in different natural zones.

1.2.4 Discussion and conclusions

The foregoing is a brief review of some of the SOM models in existence today. The total number of SOM models is very large; there are 18 SOM models in the Register of Agro-Ecosystem Models (Plentinger and Penning de Vries, 1996), and, additionally, a number of ecosystem models which consider the decomposition of organic residues. There is rapid progress in this fast-growing field of soil science and ecology, enabling quantitative evaluation and prediction of SOM behaviour in conditions of traditional land use and of climatic change (Post et al., 1996).

The first attempts at SOM modelling were sometimes connected with *theoretical* pedology. Today the models are justified by serious experimental validation, comprising short-term laboratory and field trials and long-term field experiments (Powlson et al., 1996). An attempt at development of a unified theory of soil carbon and nitrogen dynamics has been made by Ågren and Bosatta (1996).

The majority of models are based on microbiological concepts of organic matter mineralization, especially humification, without consideration of the role of the soil fauna. This is a serious drawback, because the fermentation systems of the soil fauna are thought to be the main driving force of the humification process in soil, as is well-known in forest pedology (Wilde, 1958; Duchaufour, 1961; Chertov, 1981). There are only a few recent models which consider faunal activity and the food-web as main factors affecting SOM transformation (Hunt et al., 1987; Moran et al, 1988; De Ruiter and Van Faassen, 1994; Zheng et al., 1997), and they consider fauna activity in relation to SOM mineralization mostly in terms of taxonomic functional groups of organisms, without any concept of communities of soil decomposers. De Ruiter and Van Faassen (1994) include a very clear comparison of traditional SOM models with food-web models, but fauna models do not simulate all SOM dynamics and in general only nitrogen dynamics are considered.

Including the soil fauna explicitly in a SOM model allows the influence of external impacts, primarily pollution, to be taken into account in a very simple way. It was shown using the SOMM model that it is sufficient to change the decomposing activities of the soil fauna to account for global changes of carbon and nitrogen in relation to changing conditions. Sometimes it is enough to switch off the activity of a particular group of decomposers (earthworms, for instance) to simulate perturbation of the ecosystem.

The authors of agricultural SOM models also avoid biochemical identification of the SOM fractions as humic and fulvic acids. The role of humic substances as a main product of organic residue transformation in a soil has probably been underestimated. Moreover, efforts have been made towards physical separation of SOM, for validation of the fractions used in different models (Hassink, 1994). At the moment there is really only one theoretical model (Morozov and Samoilova, 1993) with an attempt of mathematical formalization of the dynamics of humic acid. This underestimation of the role of humic substances derives from the absence of correlation between existing concepts about humic/fulvic acids and the kinetics of SOM transformation (Popov and Chertov, 1993), but a new hypothesis has still to be elaborated. On the basis of the discussion of the models, it seems to us that this should be of a functional type, based on the division of SOM into active (labile) and passive (stable) fractions. Moreover, this underestimation of humification as a specific pedological process leads to the lack of recognition of humic substances as an important SOM component influencing the kinetics of its transformation: practically all the models treat lignin and nitrogen contents only as the main agents controlling litter and SOM transformations. The SOMM model (Chertov and Komarov, 1997b) is the only exception, because of its consideration of the inhibiting role of humified matter in the process of mineralization.

For a complete ecological model of the soil compartment of the forest ecosystem, it will be necessary to develop not only a model of SOM transformation, but also a model of transformation of mineral soil. SOM and soil minerals differ in the rates of their transformation, but the mineral component influences the rates of organic matter dynamics. This additional aspect, which seems to be very important, requires that the rates of SOM transformation and their dependence on the current concentration of elements in the decomposing material should be taken into account (Ågren and Bosatta, 1987, 1996). This makes the nature of the model more complicated, as the model becomes nonlinear, but might be necessary in order to account for some important effects. These authors have made the first successful contribution to a theory of SOM mineralisation (Ågren and Bosatta, 1996).

In conclusion, we should like to emphasize that there are excellent prospects in relation to the development of SOM models. Moreover there is an understanding that SOM modelling is crucial for the development of models of forest and agricultural ecosystems. Thus practically all the recent SOM models have been elaborated in relation to ecosystem modelling. The process of SOM model evaluation has being started now (Powlson et al., 1996; Smith et al., 1997; Paustian et al., 1997). The SOM criterion of forest sustainability has been recently proposed also (Kimmins, 1990; Morris et al., 1997). We think the most productive way forward is as follows: *(i)* to include in the models new facts and ideas on SOM dynamics (Feller and Beare, 1998; Zech et al., 1997); for example, the development and use of concepts about the trophic functions of SOM, for example the supply of higher plants with fragments of organic molecules required for lignin synthesis (Popov and Chertov, 1993, 1996); *(ii)* to reflect the role of humic substances as a source of SOM formation and decomposition; and *(iii)* to make a compilation of ecosystem-oriented models, because autotrophic organisms are the main source of energy and matter for SOM transformation and soil development as a whole.

However, there are now only a few models which can be effectively used for forest ecosystem simulation. These are the LINKAGE model of Pastor and Post (1985); the model of Ågren and Bosatta (1987); the ITE model (Thornley and Cannell, 1992); the SOMM model (Chertov and Komarov, 1995a, 1997b); the ROTHC model (Coleman and Jenkinson, 1995); which has been used by Post et al. (1996) for calculation of the carbon pools in all ecosystems of the world, and the modification of CENTURY model (Friend et al., 1997). It is a fact that forest soils always have a superficial organic layer on the forest floor that influences SOM dynamics and plant nutrition. However, only two models (Ågren and Bosatta, and SOMM) include the SOM forest floor components. Friend et al. (1997) also modified CENTURY model to use it for forest simulation by including above ground SOM compartments.

1.3 Simulation Models of Forest Dynamics

1.3.1 Introduction

Simulation modelling of forest stands, populations and ecosystems is now undergoing very intensive development in the West. For example, a modification of D. Botkin's classical model, JABOWA, has recently been published (Botkin, 1993). A set of H. Shugart's models shows consistent progress in the "gap" approach (Shugart, 1984; Shugart et al., 1992a). European achievements in the field have recently been reflected in special issues of Forest Ecology and Management and of Ecological Modelling (Mohren and Kienast, 1991; Breckling and Müller, 1994). These publications demonstrate that the methodology of forest simulation is rapidly changing from numerous different approaches to a combination of individual-based and process-based models.

Forest simulation modelling has also been developing in East Europe (in the countries of the former Soviet Union) during the last few decades but the results are not well-known around the world. Interesting ideas and approaches have been developed in process-based and in individual-based modelling. The concept of a discrete description of the ontogeny of forest species based on tree morphology and physiological stage of development has been incorporated into the models. The main feature of East European forest modelling is that it has developed simultaneously with, and under the strong influence of, theoretical (analytical) modelling.

1.3.2 Classification of models

The validity of the classification of modelling approaches is debatable. Nevertheless we have tried to arrange existing types of modelling approaches as shown in Figure 1.

The types of models in Figure 1 are arranged from left to right in relation to their scale, i.e. from more detailed to more generalised. Tree models express the growth of a single tree with different degrees of detail of the processes. Individual-based models consider the growth of every tree in a forest population, taking into account competitive interactions or resources redistribution within a stand. This approach is now being developed intensively. Stand models do not operate with single trees; they simulate dynamics of numerous stands or ecosystems with generalised characteristics, sometimes with very detailed consideration of ecophysiological processes.

Combination (or hybrid) models symbolise a transition to a new level of simulation, unifying different types of models. It seems to us that combination models are the most promising area of forest simulation development. Regional models are tools for generalised simulation of forest dynamics on landscape, regional, and national scales. This type of model has a quite specific methodological background, being very different from the stand scale.

Figure 1. A classification of existing simulation models of forest dynamics. The items marked in bold are thought to be promising approaches for development in the near future.

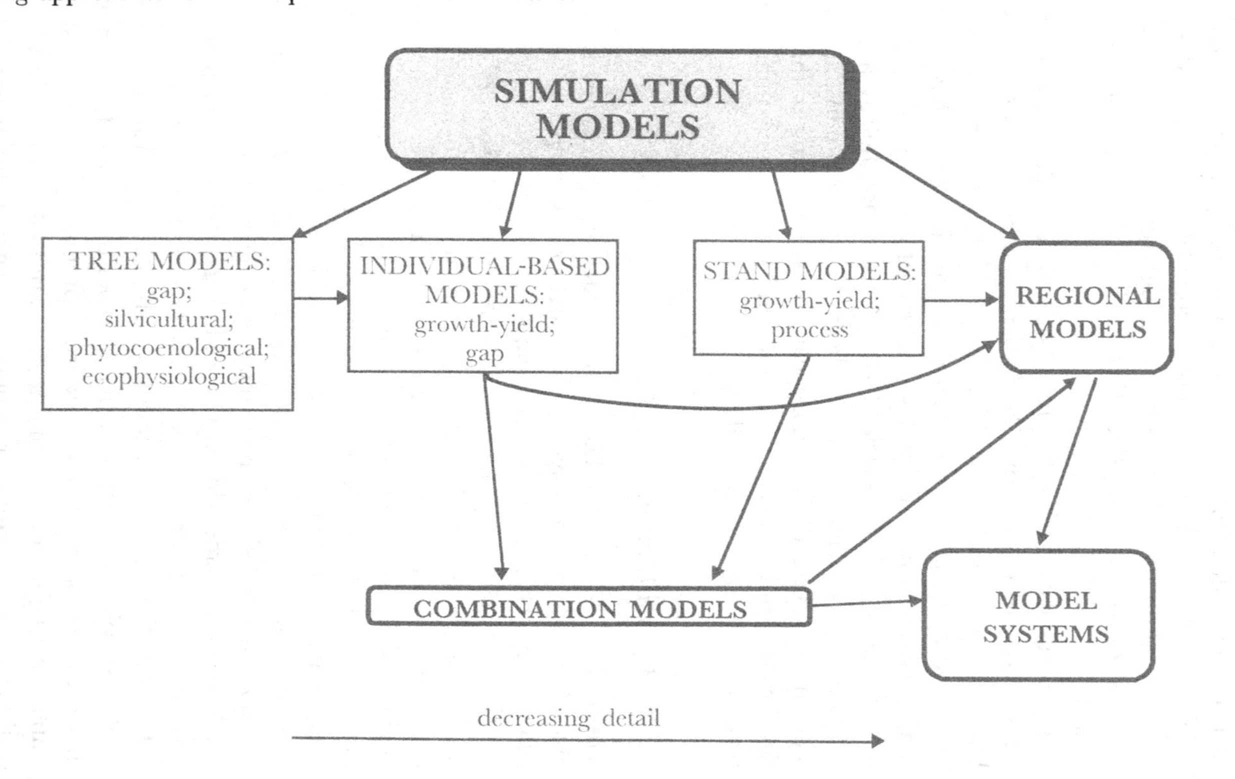

Model systems are supposed to be the next step in forest modelling. The system must be on a regional or national scale, unifying silvicultural, ecological, environmental, technological, and socio-economic models. The first manifestations of the systems approach are now evident.

Other ways of classifying modelling approaches are possible, but we think that the classification in Figure1 is logically consistent. A discussion of all the approaches follows: Table 2 contains a list of models discussed, classified according to our proposed scheme.

1.3.3 Single tree models

Modelling individual tree growth using dendrometrical data began in the last century. The problem was previously discussed in detail by forest scientists (Kuzmichev, 1977; Terskov and Terskova, 1980; Kiviste, 1988) and it has also been under consideration recently (Alekseev, 1993). These models always represent functions of tree growth in a stand; the total number of the functions sometimes reaches one hundred. This approach can be classified as "silvicultural models of tree growth".

The models of Scots pine and Norway spruce of Pukkala (1989b; Pukkala et al., 1994; Miina et al., 1991) describe tree diameter and height increment in relation to stand characteristics and an index of competition (considering distance to and diameter of nearest neighbouring trees). A Scots pine model by Kazimirov and Mitrukov (1978) unifies both dendrometrical and biometrical approaches, with functions predicting diameter, height and biomass of different tree parts in relation to age, stand characteristics and some site conditions. These models are based on very comprehensive experimental data and they are applicable only for defined bioclimatic conditions. They are expressed as complicated regressions between the variables and are not formally dynamic models.

The other type of tree growth simulator represents a kind of model first elaborated two decades ago by Botkin et al. (1972) and then intensively used for gap simulation models by Shugart (1984; Shugart et al., 1992a) and other researchers (Mohren and Kienast, 1991; Breckling and Müller, 1994). These models comprise one equation for DBH increment in relation to some species-specific characteristics and light conditions in a stand. The number of these characteristics is generally about twenty (e.g. Prentice and Helmisaari, 1991) and includes the maximum age, DBH, height and height increment

Table 2. List of tree and forest models discussed in the review, in chronological order.

The authors of the models	Single tree models				Individual-based models		Stand Models		Combi-nation models	Regional models	Models' Application		
	Silvi-cultural	Gap	Phyto-coeno-logical	Eco-physio-logical	Growth-yield	Gap	Growth-yield	Process			The-oretical	Silvi-cul-tural	Envi-ron-mental
1	2	3	4	5	6	7	8	9	10	11	12	13	14
Botkin et al., 1972		X				X					X	X	
Kazimirov and	X											X	
Mitrukov, 1978													
Rachko, 1978				X							X		
Hari and													
Kellomäki, 1981				X							X		
Chestnykh, 1982								X			X		
Kull and Oja, 1984				X							X		
Moshkalev et al.,							X					X	
1984													X
Pastor and Post, 1985								X			X		
Arp and Mc Grath,						X					X		
1987					X						X	X	
Kienast, 1987			X								X		
Pegov, 1987		X				X						X	
Bogatyrev, 1988													
Keane et al., 1989													
Kull and Kull, 1989				X							X		
Pukkala, 1989	X										X	X	
Korzukhin, 1989 Oja,								X			X		
1989												X	
Chertov, 1990			X						X		X		X
Chertov et al., 1990										X			X
Krieger et al., 1990				X				X			X		X
Mäkelä, 1990				X				X			X	X	
Post and Pastor, 1990		X				X			X		X		X

Table 2. cont.

1	2	3	4	5	6	7	8	9	10	11	12	13	14
Prentice and Leemans, 1990		X				X					X		
Frelich and Lorimer, 1991										X	X		
Miina et al., 1991					X							X	
Mohren et al., 1991									X				X
Urban et al, 1991		X				X						X	
Alexandrov, 1992										X	X		
Arp and Oja, 1992				X				X			X		X
Chumachenko, 1992		X				X					X		
Kellomäki et al.,				X		X		X	X		X	X	
1992										X	X		
Knyazkov et al., 1992		X				X					X		X
Shugart and Prentice, 1992													
Botkin, 1993		X				X		X	X		X	X	X
DeVasconcelos and Zeigler, 1993										X		X	X
Schuster et al., 1993										X		X	X
Pacala et al., 1993		X				X					X		
Bossel, 1994								X			X		X
Kirschbaum et al., 1994								X					X
Mohren, 1994								X			X		X
Pukkala et al., 1994	X										X	X	
Bevins et al., 1995		X				X		X	X		X		
Chertov and			X					X		X		X	
Komarov, 1995,													
1997		X				X		X	X		X		X
Bugmann, 1996													
Chumachenko et al., 1996		X				X				X		X	
Lindner et al., 1996										X			X
Friend et al., 1997				X		X		X	X		X		X

* We used authors names only because great deal of the models have no acronyms.
The characteristics of combination models include types of models composing the "combination".

of the trees in the stand. Climatic, shade and water stress tolerance are also of major importance but site quality is considered only in some of these models.

Neither silvicultural nor gap models account for the organic matter balance of a tree; thus they cannot be used for simulation of the biological cycle in forest ecosystems. Perhaps the next stage in tree models of the gap type is represented by the models of 30 woody species in the ForClim model (Bugmann, 1996), which is really a combination model. The species are in five groups in relation to light conditions and nitrogen response. There are functions expressing the response of the trees to degree-days for the vegetation period and to evapotranspiration deficit as an expression of drought conditions.

Another type of tree model is that proposed by the authors, with the name "phytocoenological" model (Chertov, 1983a; 1990; Chertov and Komarov, 1995a). It is a representation of biomass growth for calculation of carbon balance in the ecosystem as a difference between total biomass increment and tree litter, using the classic equation of Bertalanfy (1942), but without consideration of photosynthesis itself. The total increment, as the simplest expression of net photosynthesis, is calculated using the silvicultural concept of the biological productivity of leaves or needles (Kazimirov and Morozova, 1973; Kazimirov et al., 1977, 1978). The maximum value of this variable is dependent on climate and is species-specific. Its value for a tree in a stand is controlled by light, nitrogen and water supply. A set of other species-specific ecological parameters is also necessary for effective use of the model (Chertov, 1983b; 1990; Chertov and Komarov, 1995a). For example, it is the specific consumption of nutrient required for the synthesis of a unit of biomass increment which is a quantitative expression of edaphic plant types (oligotrophic, mesotrophic, eutrophic, etc.). The last two model types allow calculation of both biomass dynamics and dendrometrical characteristics of a growing tree. A semi-analytical tree model of Bogatyrev et al. (1988), and an early gap model of Reed (1980), are close to this description.

Eco-physiological models of tree biomass growth, considering photosynthesis, respiration, assimilates translocation, and litter formation, are more complicated and comprehensive. The cores of these models are sub-models of photosynthesis, respiration and assimilates translocation (Rachko, 1978, 1979; Hari and Kellomäki, 1981; Kull and Oja, 1984; Kull and Kull, 1989; Mäkelä, 1990; Friend et al., 1997), which define relatively simple approximations of the dependence of the processes on environmental variables and available resources, particularly nitrogen. Some models include the concept of L-systems, as proposed by Poletaev (1966), for the mathematical description of Liebig's bottle-neck principle (Rachko, 1978; Kull and Oja, 1984; Kull and Kull, 1989).

1.3.4 Individual-based models

The recent review by Liu and Ashton (1995) contains a very reasonable classification of individual-based models of forest dynamics. The authors divide the models into two groups: growth-yield models and gap models. They mention that growth-yield models are used by foresters to assist timber production and evaluate growth and yield of one to several commercial timber species in managed forests, while gap models are usually developed by ecologists to explore ecological mechanisms and patterns of structure and functional dynamics in natural forest ecosystems (*ibid*, p. 157). Growth-yield models use regression equations for quite precise description of growth of every tree in even-aged managed stands, depending on site-specific environmental and species variables. Gap models simulate development of forests through the process of natural regeneration and formation of uneven-aged stands and taking into account light availability for every tree.

1.3.4.1 Growth-yield individual-based models

This approach arose as a generalisation of experimental data on stand growth observed on permanent sample plots (Liu and Ashton, 1995). This approach permits precise prediction of tree height and diameter in even-aged, mostly monodominant stands; the process of regeneration is thus not considered in the models. The specific property of the models is their spatial structure with simulation of tree displacement. This allows analysis of the role of spatial structure in stand development. An example of such a model is that of Miina et al. (1991), which describes the role of peatland drainage on growth and spatial characteristics of Scots pine stands. Another example is a Russian model by Pegov (1987), for growth and horizontal structure of silver birch stands in north-west Russia. The model calculates the development of a pure stand in which self-thinning is caused by light reduction or by stochastic mortality. The author used Monte Carlo simulation with different initial random values for initial tree parameters having the same mean values.

1.3.4.2 Individual-based gap models

This type of forest simulation model is currently more popular. The results of application of this approach (the JABOWA-FORET model type) are well-known, having been published in a large number of reports and books. A list of the models is included in reviews by Shugart et al. (1992a) and Liu and Ashton (1995). The approach has stimulated the development of forest succession theory and the models predict the response of forest ecosystems to forest management practice, forest fire, industrial pollution, and climate change (Shugart, 1984; Kienast, 1987; Keane et al., 1989; Mohren and Kienast, 1991; Shugart et al., 1992b; Botkin, 1993; Breckling and Müller, 1994).

The models utilise tree models of the gap type discussed above, with probabilistic procedures for the mortality of suppressed trees and special routines for gap formation and seedling establishment. The approach has been applied to simulation of temperate, boreal and subtropical forests. A number of models have originated from the early models of Botkin and Shugart, with practically the same structure, but the existing models have sometimes been applied to new conditions; for example, the ZELIG model of Smith and Urban (1988) was applied in The Netherlands (Mohren et al., 1991).

A very successful attempt at gap modelling was made in Estonia by Oja (1989). His model SJABO, of the JABOWA type, has been created for Estonian mixed conifer-broadleaf forests. The model has quite detailed consideration of all tree light conditions and of soil water and fertility, with calculation of tree growth with regard to Hegyi's competition index.

The only Russian model of the gap type (Chumachenko, 1992) has been created for Central Russian broadleaved uneven-aged forests. It has some interesting features: (i) the space is represented as a discrete isometric system with the size of a 3-D pixel, 0.5 × 0.5 × 1 m, and every tree crown is described as a set of pixels; (ii) there is detailed consideration of tree ontogeny in relation to experimental data for every species, with different tree behaviour for every age stage and regeneration pattern (Smirnova et al., 1990); and (iii) the treatment of light penetration and shading at the simulated geographical point is very precisely dependent on the disposition of neighbouring trees, taking into account the trajectory of the sun.

Nowadays, there are some variants of the individual-based approach. The FORSKA model (Prentice and Leemans, 1990; Leemans, 1992) has a structure in which one patch is the regular spatial unit with a set of Monte-Carlo simulations. However this model operate not with individual trees but with tree age cohorts only. The other approach is currently being developed by Chumachenko et al. (1996), combining calculation of the light conditions in vertical layers in each patch, taking into account the growth of every tree. The SORTIE model of Pacala et al. (1993) is a development of the approach by making use of comprehensive experimental material as a basis. It is the only model which has statistically significant species parameters and a procedure by which light reduction is obtained from direct measurements of light interception in the forest. SORTIE's authors criticise other modellers who make only little use of experimental data.

We can conclude that the gap approach is very fruitful, especially when linked to the large predictive and theoretical significance of forest succession. However, it seems to us that the existing gap models cannot solve all the problems of forest succession, because: (i), they do not consider site specific forest development in every study area; and (ii), they do not take account of tree-soil interactions and biological productivity. Gap models treat soil as an external factor or as an additional compartment for nutrient supply, but without consideration of soil development as a component of the forest ecosystem. They are, therefore, unable to simulate primary ecogenetical succession with simultaneous development of the forest stand and its soil. Gap models really simulate secondary succession alone, assuming constancy of edaphic conditions. There are many experimental data showing that this idea is misconceived, because there are well-proven significant soil changes in the forest ecosystem, even within one forest generation (Odum, 1971; Chertov, 1981; Razumovsky, 1981; Sennov, 1984).

1.3.5 Stand models

1.3.5.1 Growth-yield stand models

Chronologically, these models were developed earlier than the individual-based ones, but logically they come next. They represent a correct mathematical approximation of existing yield tables and site index curves. They are really a set of regression equations supplementing yield tables (Moshkalev et al., 1984). Nevertheless, there have been successive efforts to develop this approach for predicting stand response to forest fertilisation and thinning (Hynynen, 1995). A large pool of regression equations describing dependence of stand growth on climate, soil moisture, and chemical properties (see review in Chertov, 1981; Kazimirov et al., 1985) also belong to this type of model. A dynamic model of stand was also developed by Karev and Skomorovsky (1998). This model is of the analytical type and will be discussed in Chapter 3.

1.3.5.2 Process models

Process simulation models are actually also of the stand model type, because they describe formation of stand biomass, treating eco-physiological processes in the tree canopy without considering separate trees. They are of three kinds: the first considers only eco-physiological processes. The second considers eco-physiological processes of biomass synthesis and also soil processes of organic matter decomposition, allowing simulation of the organic matter balance (i.e. the biological cycle) in the forest ecosystem. The third deals with biogeochemical soil processes mostly. An example of the first kind is the model by Korzukhin et al. (1989), which is combined with a population model.

However, models of the second kind are now dominant. They perhaps originated from the FORTNITE model of carbon and nitrogen dynamics in forest ecosystems (Aber and Melillo, 1982) and the well-known LINKAGES model of Pastor and Post (1985). This has a soil sub-model as an independent unit (see section 1.2, on models of soil organic matter) and a production module. A number of comprehensive process models has been developed (Krieger et al., 1990; Arp and Oja, 1992; Kirshbaum et al., 1994; Bossel, 1994; Mohren, 1994) and there has recently been a extensive review of 16 process models (Tiktak and Van Grinsven, 1995). The tree canopy productivity routines in these models mostly correspond to those being elaborated for eco-physiological tree models. Soil routines are usually quite simple SOM mineralization sub-models concentrating mostly on nutrient release, although there are some models with detailed consideration of soil chemistry and mineral weathering. The models, therefore, allow simulation of effects of pollution and water stress on forest ecosystems. The model of Chestnykh (1986) is an example of a Russian process model, which takes into account daily photosynthesis, respiration, assimilate translocation, and soil organic matter decomposition, for a Norway spruce stand in the southern taiga.

The modelling approach of Arp and McGrath (1987; Arp et al., 1987) is intermediate between process and stand modelling. Their model is a set of models of two, four, and eight compartments, with a very simple description of stand biomass increment and litter mineralization, for a fixed geographical point with particular environmental conditions.

The third class of models are also discussed in review of Tiktak and Van Grinsven (1995) and can be represented by Ågren and Bosatta (1996) genera of models based on their soil decomposition theory, and the SAFE model of biogeochemical processes (Warfvinge and Sverdrup, 1992; Sverdrup et al., 1995). SAFE is the model of soil cation exchange capacity and a weathering of soil minerals as influenced by natural processes and atmospheric acidification. The models accentuate soil biochemistry processes in forest ecosystems and are joined with forest watershed models.

1.3.6 Whole ecosystem combination (combined or hybrid) models

The necessity to combine models of different types, first of all, gap and process models, is obvious and has been stated many times (e.g. Dale et al., 1985; Huston et al., 1988; Bossel, 1991; Shugart et al., 1992a,b; Levine et al., 1993). This is a model type now in the process of being intensively developed. It allows a comprehensive analysis of forest ecosystem dynamics. A well-known FORCYTE model (Kimmins and Scoullar, 1979) is an excellent prototype of recent models of this type, being a combination of a simple stand growth simulator with SOM decomposition model. FORCYTE is the first model having a user-friendly interface for education and professionals. Now there is already a successful attempt by Post and Pastor (1990) to develop their LINKAGES process model as an individual-based model. The modified model has demonstrated that stand-initiated soil changes during succession influence stand dynamics and productivity. The recent version of the classical JABOWA model is also of the combination type (Botkin, 1993).

The SIMA and FinFor models of Kellomäki et al. (1992; 1993) are other prototypes of combination ecosystem models. They operate with tree age cohorts simulator combined with SOM model by Pastor and Post (1985). Bevins et al. (1995) have recently described the structure of their combination models in detail. The ForClim model is also of this type (Bugmann, 1996) and was specifically created to quantify the effects of climate on species composition and productivity in

Central Europe. This model simulates population dynamics as a gap model, soil processes using the LINKAGES model (Pastor and Post, 1985), and has an additional climatic component for generation of weather conditions. However, the model has no nitrogen inputs from the atmosphere and takes no account of carbon and nitrogen balance in the ecosystem.

The advantage of combination approach is demonstrated by the Hybrid model (Friend et al. 1997). It is an individual-based tree-soil spatial model. The model have a detailed eco-physiological sub model of plant photosynthesis, respiration and carbon translocation, and a modification of CENTURY model of SOM decomposition (Parton et al., 1988, 1993). Additionally it has grass layer and soil hydrology sub models and climate generator. The model simulates plant growth and soil processes at daily time step with realistic patterns of stand and soil dynamics. However the current version of the model represents a theoretical tool because it operates with three generalised functional plant types only (C3 grass, cold deciduous broadleaved and evergreen needleleaved trees).

The authors' model EFIMOD is a combination model of this type (Chertov and Komarov, 1995b, 1997b; Chertov et al., 1998a, 1998c). It is based on the concept of a single plant ecosystem, SPE, (Chertov 1983a, 1990) in which a single higher plant occupies a certain space in the above-ground environment and in the soil. The SPE can be treated as an elementary cell of a forest ecosystem, and the ecosystem itself can be represented as a group of SPEs, in which the trees are located in "personal soil pots". Pot size (area of nutrition) depends on the root mass and the rate of nutrient consumption and changes continuously as the tree grows, reflecting the competition for soil nutrients and water. The SPE concept unites Uranov's (1965) ideas of the "phytogenic field" (an area strongly influenced and controlled by a plant) and Karpachevsky's (1981) ideas of the "tessera" (the soil associated with an individual tree). Moreover, the recently formulated "ecological field theory" (Wu et al., 1985; Mou et al., 1993; see also Chapter 2 of this review) is very close to the SPE concept. EFIMOD simulates biological productivity of the tree and the soil processes of every SPE. So the basic units of the model are SPE modules unifying the tree with SOM modules. EFIMOD comprises the tree module described above and the soil module based on the authors' SOMM (see Section 1.2). The first version of the model demonstrated the high response of the soil component to changes of stand characteristics

(Chertov, 1990). The tests of the recent version showed high sensitivity of stand and soil to tree nitrogen response, climate and soil with their quite complicated interactions, nitrogen input from the atmosphere, stand density and pollution (Chertov et al., 1998a, 1998b, 1998c).

We can conclude that combination, whole ecosystem models will allow the simulation of all types of forest succession, primarily ecogenesis (primary succession), with simultaneous development of vegetation and soil. This was not practically possible when using separately gap and process models. These possibilities of combined models are still not fully appreciated in the West, although they are considered to be a major advantage of this new approach. Theoretically, this type of model is also very suitable for simulation of biodiversity in forest ecosystems. However, these models cannot completely solve the problem of forest succession, as discussed in the next section.

There are two problems in this field which are currently not solved in combination approach: modelling ground vegetation, and the ecological parameters of the trees (silvics).

It is well-known that ground vegetation uses up a large part of the available nutrients and water in forest ecosystems. However, modelling the ground vegetation is still not well-developed in forest simulations. Perhaps COVER was the first attempt to simulate ground vegetation and shrub layers in a forest (Moeur, 1985). SIMA is one model which has a well-developed sub-model of ground vegetation (Kellomäki et al., 1992). This sub-model simulates growth and litter fall of three groups of ground vegetation (pioneer, intermediate, and climax) on mesic sites. Plant growth in the model is dependent on external variables, particularly temperature, precipitation and light conditions. There is also a model of post-fire succession of heathland dwarf shrubs by Van Tongeren and Prentice (1986), which allows simulation of the spatial population dynamics of three species. The functional sub model of "C3 grass" is also in the Hybrid model of Friend et al. (1997). This is really a problem to be solved in the near future by incorporating ideas and methodology of population and process approaches developed in grassland modelling (Komarov, 1988; Parton et al. 1988; 1993).

Defining the ecological parameters of trees is a problem of great importance. Different kinds of species parameters have already been used in the simulations that exist. Botkin (1993) used "species site qualities" and "species response functions". Reed (1980) writes about "growth response". Shugart et al. (1981) utilise the same parameters ("case-specific response") and Kellomäki et al. (1992) use "growth multipliers". Oja (1989) and Prentice and Helmisaari (1991) reported on the silvics of about 20 European species for models of the gap type. We generalised the experimental data needed for the ecological parameters in our model for the main tree species of European boreal forests (Chertov, 1983b; Chertov and Komarov, 1995b). Almost none of the authors emphasizes the methodology and validation of the parameters used; only the model of Pacala et al. (1993) has parameters well validated with experimental data. We consider that definition of the ecological parameters of species will very soon be a bottle neck in the development of combination tree-soil models with elements turnover. Our opinion is that elaboration of a set of ecological parameters and functions for individual tree species is now a specific and important task. The parameters will be a quantitative expression of the main autecological and ontogenetic patterns of the species. Bugmann's (1996) proposal of "tree functional types" has been considered to be a temporary compromise solution. A special problem in this field is the absence of models in which the ontogenic phases of trees are taken into account. Chumachenko's (1992) model is the only attempt to utilise ideas and data on tree ontogeny. This is a serious drawback in simulation approaches, so we have included a special section below, elucidating the problem in more detail.

1.3.7 The discrete description of the ontogeny of plants and trees

We have previously discussed individual-based tree models with continuous variables, that describe a tree located at random points and interacting with its surroundings. The behaviour of the population has been described by a set of parameters and variables corresponding to each tree. It is apparent that if the number of trees in the population is large then the contribution of each individual tree is relatively small. The question then arises as to whether we could find a simpler way of describing growth of the individual tree, taking into account the large size of the population. Is it possible to describe the growth and development of a tree as a discrete sequence of developmental stages, considering that the development of the tree will be the same at the same ontogenetic stage? The continuous growth and development of the tree can be described as a discrete chain of states.

Therefore each tree can be expressed as an automaton with a finite set of states. It is fortunate that the biological background for this kind of plant description has been developed independently of the problems of mathematical modelling.

Plant or tree ontogeny may be characterized not only by physical time (years, months, etc.), but also by biological time inherent in the life of the organism (Harper, 1977). Division of the continuous ontogenetic process into functional stages allows the coherent scales of biological time for different species with different life spans to be defined. This approach has been developed in the work of Russian botanists, as the concept of discrete ontogeny (Rabotnov, 1950, 1978; Uranov, 1975; Gatzuk et al., 1980; Serebryakova, 1977; White, 1985; Zaugolnova et al., 1988).

A detailed study of the ontogeny of a large number of species in different communities allows us to state the main principles of this concept (Smirnova et al., 1998):

1. The continuous process of individual plant development may be subdivided into several stages, on the basis of structural indicators which reflect functional importance. These are: the presence or absence of embryonic, juvenile, or mature morphological features; the ability of an individual to reproduce or to propagate vegetatively; the ratio between alive and dead, and between growing and non-growing plant parts.

2. Ontogenetic stages are universal and apply to plants of various life forms. The ontogeny of a plant can be subdivided into four periods and nine stages.
   - I. Latent period (seed).
   - II. Pre-reproductive period (seedling, juvenile, immature, semi-mature).
   - III. Reproductive period (young, mature, and old-reproductive).
   - IV. Post-reproductive (senile) period.

3. Individual development may be expressed in different ways (polyvariant development). Polyvariant development means both temporal and structural diversity of ontogenetic pathways. Temporal variations are expressed as:
   - a) developmental delays, i.e. longer duration of certain ontogenetic stages in unfavourable conditions as compared with the duration of the same stages in optimum conditions

b) omission of some ontogenetic stages
c) recurrence of earlier ontogenetic stages.
Structural variations are expressed as:
a) different vitality
b) different life forms
c) different types of propagation.

4. There is no direct correlation between ontogenetic stage and calendar age.

We shall consider several properties of the discrete description of the ontogenetic stages of trees in more detail, bearing in mind their applications in forest modelling.

The sequence of ontogenetic stages of a tree is shown in Figure 2 The distinction between ontogenetic stage and age (Gatzuk et al., 1980) was made on the basis of morphological and ecological studies.

The ontogenetic stages of trees have the following main features (Serebriakova, 1977; Zaugolnova et al., 1988; Smirnova et al., 1998):

Seed (se) is characterized by size and biomass, degree of embryo development, duration and type of dormancy, nutrient storage.

Seedling (pl) usually possesses a partially heterotrophic nutrition type, i.e. it uses both the substances of the maternal plant stored in the seed and its own assimilates. It usually consists of a primary shoot and a primary root.

Juvenile plant (j) is usually simple in structure. A juvenile tree has a small unbranched primary shoot bearing juvenile leaves or needles. Its root system consists of a primary root and a few lateral ones.

Immature plant (im) has a structure which is transitional between juvenile and mature. A plant begins to branch at this stage, so its shoot system consists of branches of a low order. The leaves or needles display mature form and structure; exceptions are the species with compound leaves. The root system includes either the entire primary root or its remains, lateral roots of the second and higher orders and, for some species, adventitious roots.

Semi-mature plant (sm) has mainly mature features. It is a young tree with a distinct trunk and crown. It does not produce seeds, but the leader shoot has maximal annual increment. The tree has a crown of elongated shape with a distinct leader axis. This stage is of fundamental importance, because the light demand increases sharply.

Young reproductive plant (g_1) is similar to the adult tree. The reproductive structures appear; the seeds are located within the upper part of crown, but the quantity of seeds is small. Height growth is rapid. Trees in the ontogenetic stages from *j* to g_1 are generally in a phase of exponential growth.

Mature reproductive plant (g_2) has a decreasing rate of vertical growth, while the radial increment increases to a maximum. The crown and root system attain their maximal size and branching order. The bark becomes rougher. The number of seeds is maximal.

Old reproductive plant (g_3) has cessation of height growth; radial increment is very small. The size of both crown and root system decreases because many frame branches and anchor roots have already died off. A secondary crown may in some cases replace the primary one. Tree biomass is still increasing as a result of the continuous increase in trunk diameter. Seeds are produced irregularly.

Senile plant (s) has live shoots only in the secondary crown, and the leaves may be of the juvenile type. The upper parts of the crown and trunk are lost; the root system is declining. No seeds are produced. Trees of g_3- and *s*-stages show suppressed growth.

Size differences may also be expressed as vitality levels. The vitality level may change in each ontogenetic stage if the conditions become unfavourable. Trees of all vitality levels were found in many natural forests. We shall briefly describe the main features of three vitality levels (Fig. 2, Smirnova et al., 1998).

(1) *Normal vitality*. This usually occurs in optimal conditions. The plant has maximum size, biomass, and longevity, relative to other trees of the same ontogenetic stage growing in other habitats. A tree of normal vitality grows and develops without delays, omissions or recurrences of ontogenetic stages. Size and biomass increments are strongly correlated with calendar age. Quantitative traits of vitality depend on the geographical region.

(2) *Sub-normal vitality*. The plant has smaller size and biomass than normal. The pre-reproductive period (*pl, j, im, sm*) is longer; reproductive (g_1, g_2, g_3) and post-reproductive (*s*) periods are shorter, than in a plant with normal vitality.

(3) *Low vitality*. The plant has minimal size and biomass, and it grows very slowly. The top of the tree frequently dies off, and the crown has many dead branches. The ontogeny may be incomplete because some ontogenetic stages are omitted or the latest stages are absent. The omission and recurrence of ontogenetic stages are connected with the death of above-ground parts of the tree.

Figure 2. General scheme of the ontogenetic stages of a tree (Smirnova et al., 1998). 1 – normal vitality, 2 – sub normal vitality, 3 – low vitality. p – seedlings, j – juvenile, im – immature, sm – semi-mature, g_1 – young reproductive, g_2 – mature reproductive, g_3 – old reproductive, qs – quasi-senile, s – senile stages.

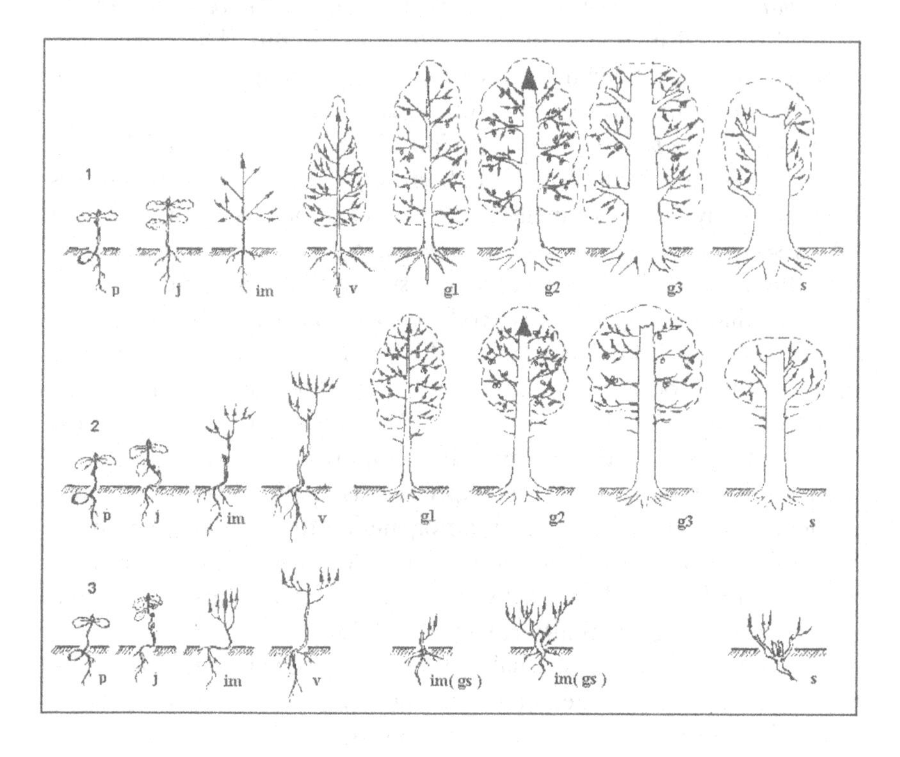

The ontogenetic concept proposed by Rabotnov has been verified by investigations of species with different life forms. The ontogenetic analysis of trees shows different variants of ontogenetic pathways for trees growing under diverse coenotic and environmental conditions. Individuals of the same species can have different ontogenetic patterns, even within the same community, as a result of belonging to different elements of the population mosaic within natural forests. Individuals of different species show high individuality of ontogenetic behaviour; this is especially clear in the heterogeneous natural forest community. A study of the ontogeny of a species throughout its geographical area reveals a spectrum of ontogenetic polyvariancy for every species that determines the potential of the species. In any one community only part of this potential is realized. We take these results into account as a basis for forecastig species behaviour in the permanently changing environmental mosaic of the natural community.

Descriptions of age status exist for more than 400 plant and tree species, for the boreal and temperate zones. However, there is a large gap in our knowledge about shifts in numerical values of physiological and ecological parameters of tree species at different age, ontogenic and vitality stages. This is a serious problem for simulation modelling, although it does not limit the use of the approach in analytical modelling.

1.3.8 Regional models of forest dynamics

This type of dynamic model is currently being widely discussed and developed. Two recent models of interest, of the Markov chain type, are a landscape scale model of age structure as influenced by a regime of storm disturbances (Frelich and Lorimer, 1991) and a model of community dynamics as influenced by fire (De Vasconcelos and Zeigler, 1993).

Our view is that dynamic simulation models, i.e. individual-based, stand and combination models at smaller scales, should be the basis for large spatial scale, regional, landscape models. An algorithm for the transition from stand to landscape scale was proposed by Urban et al. (1991) and by Shugart et al. (1992a), discussed by Flechsig et al. (1994), and realised recently by Acevedo et al. (1995). The idea is to aggregate individual-based simulations of every spatial patch into "appropriate states" in the successional sequence, i.e. for different forest sites/types, as we understand them. These simulations are the basis for parameterization of the succession, and calculation of the transition probabilities, for landscape-scale Markov models of forest dynamics for every spatial patch, using GIS technology.

We regard this methodology as correct in principle, although it actually reflects gap theory in which the role of gap dynamics is overestimated. We regard gap dynamics as of importance in native (natural, virgin) forests but we consider that large-scale disturbances (thinning, felling, fires, storms, recreation, and pollution) are driving forces of regional-scale vegetation dynamics in managed forests in all natural zones. This opinion does not reject the crucial role of the individual-based, combination approach.

In this connection, it is useful to discuss some Russian approaches. They are based on Razumovsky's (1981) concept of the succession system, which is a further development of Clements' (1916) classic theory of climatic climax and the concept of ecogenetic (endogenic, primary) succession. The succession system represents the idea of definite ranges of primary and secondary succession within groups of sites (Clements' series) in every botanical, geographical region. The concept unifies ecological, floristic and geographical data. There has been a successful attempt at mathematical formalisation of the theory (Alexandrov, 1992), and the results of modelling succession as Markov chains (Knyazkov et al., 1992) allow prediction of both community types and floristic composition. The land structure changes have been simulated in another model (Chertov et al., 1990) showing retrogressive succession as a result of land degradation. This showed a strong decrease in undegraded lands, with a corresponding increase in bad-lands, although the proportion of weakly degraded lands remained stable.

A very specific type of regional model has been proposed by Chumachenko et al. (1996), who simulated patch dynamics (16 × 16 m) for an area of forest of some thousands of hectares, using his individual-based model of the gap type. The modelling approach of Lindner et al. (1997) represents the other variety of such regional models. He used combined models FORSKA (Prentice et al., 1993) and ForClim (Bugmann, 1996) without their transformation for 10 × 10 km grid cells with an expanding of the results of stand simulation to the grid level, as we understood. The models were used to calculate steady-state parameters of forest stands and soils by long-term simulation on bare sites for recent climate and different scenario's of climate changes.

Because regional scale modelling is now just beginning, it is useful to suggest a more suitable methodology for this purpose. This is done in Figure 3 and in the text below. Figure 3 is a modified presentation of basic ideas by Urban et al. (1991, p. 108, Fig. 10) with the addition of the authors' vision of the problem. We see the following steps to be necessary for the creation of a regional (landscape) model:

1. A description of the landscape (forest sites) structure of the area, both spatial and parameterized, co-registered with soils and climate, using GIS technology.
2. A compilation of measured and/or modelled types of forest dynamics for each land unit (especially forest sites), with the rates (or probabilities) of transition. A silvicultural, ecological, pedological and environmental parameterization of every stage of succession

Figure 3. Possible methodology of modelling forest dynamics at different scales and compilation regional models of forest dynamics.

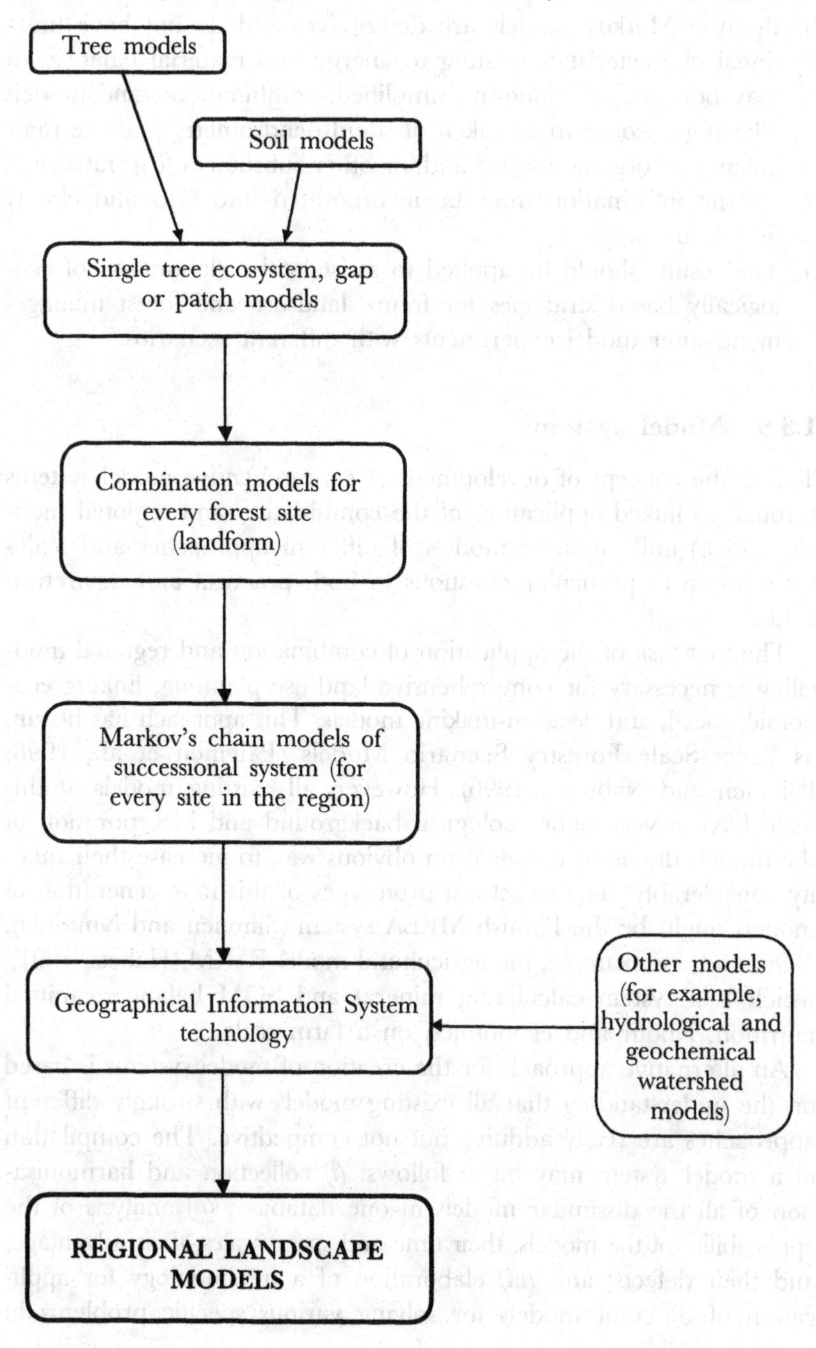

is of crucial importance. Insertion of any new factor affecting the dynamics should be possible.
3. Because Markov models are descriptive, and do not have functional characteristics relating to energy and material balances, it may be useful to elaborate simplified, combination stand models allowing account to be taken of stand dendrometry and the mass balance of organic matter and/or other substances (e.g. nitrogen).
4. All the information must be incorporated into GIS and closely linked to mapping.
5. The results should be applied to assist in the elaboration of ecologically based strategies for future land-use and forest management, after model experiments with different scenarios.

1.3.9 Model systems

This is the concept of development of comprehensive model systems through a) linked application of the combination and regional models, and b) unification of models of different approaches and scales for solution of particular questions in both practical and theoretical fields.

The first task of the application of combination and regional modelling is necessary for comprehensive land-use planning, linking economic, social, and decision-making models. This approach has begun, as Large-Scale Forestry Scenario Models (Päivinen et al., 1996; Päivinen and Nabuurs, 1996). However, all existing models at this scale have a very poor ecological background and incorporation of the models discussed earlier is an obvious way to increase their quality considerably. The structural prototypes of this new generation of models might be the Finnish MELA system (Siitonen and Nuutinen, 1996) and, for example, the agricultural model FARM (Habets, 1991), which is a system calculating mineral and SOM balances, animal nutrition, labour and economics, on a farm scale.

An alternative approach for the creation of model systems is based on the understanding that all existing models with strongly different approaches are really additive but not competitive. The compilation of a model system may be as follows: *(i)* collection and harmonisation of all the dissimilar models in one database; *(ii)* analysis of the applicability of the models, their time and space scales, their advantages and their defects; and *(iii)* elaboration of a methodology for application of different models for solving various specific problems in modern forestry practice, particularly with respect to the development of sustainable, environmental, forest management. For example, growth-yield, individual-based and stand models can be used for short-term precise prediction of stand development in managed, mostly single species, even-aged forests at short time intervals. Process and combination models can be applied to quantification of the flows of energy and matter in forest ecosystems under the impact of forestry practice and environmental stresses. Individual-based gap models will be useful for evaluation of long-term dynamics in naturally developing forests. The application of different regional models should be for regional predictions of forest cover dynamics with alternative strategies of forest management. A selection of specific models for special tasks should be an instrument for a more effective and fast prediction of forest dynamics and growth in different natural and economic conditions.

We think that such model systems should include all other types of environmental models, such as climatic (meteorological), hydrological, and biogeochemical models, and, perhaps, as yet non-existent models of biodiversity. At the same time, the systems can be a real instrument of environmental management as a whole, unifying silviculture, agriculture, water management, and all other types of land-use and exploitation of natural resources.

2 Spatial Forest Models

2.1 Introduction

Attempts to develop theoretical ecology on the basis of advances in mathematics and theoretical physics have been going on in the Soviet Union over the last three decades. The participation of leading mathematicians and physicists in this work has attracted many young scientists. It is sufficient to mention that the First All-Union Conference on Mathematical Modelling in Biology in 1975 included more than 500 original contributions. The broad spectrum of different approaches, skills and experience has led to improved understanding of fundamental processes and their hierarchy in the modelling of ecosystems. Some results from that period are described in Chapter 3.

Points of view about mathematical modelling of forest and other plant communities have changed in the last two decades. Previous models treated vegetation cover as a continuous turbid layer, sometimes stratified. The vegetation cover can then be modelled as a continuous medium using and differential equations. The available mathematical techniques required exact analytical solutions; use of numerical methods was restricted and was applied to the solution of analytical problems.This approach began in Russia in the 1950s (Khilmi, 1957, 1966) and was subsequently developed in many books and papers (e.g. Gimmelfarb et al., 1974; Ross, 1975; Svirezhev and Logofet, 1978; Bikhele et al., 1980; Korzukhin, 1980; Terskov and Terskova, 1980; Bondarenko et al., 1982; Karev and Treskov, 1982; Semevsky and Semenov, 1982; Oja, 1985a; Zaslavsky and Poluektov, 1988; Korzukhin and Semevsky, 1992). The main results were summarised in the monograph by Ross (1975). However, such models do not take into account the spatial structure of the vegetation which is composed of individuals forming the population.

Since then, new studies, in the manner typical of physicists, have established new principles regarding the modelling of plant populations, plant communities and ecosystems at a range of scales. The idea that the plant community consists of a set of separate growing individuals with local interactions among them has been confirmed as the background for spatial models of forest and other plant communities.

The main feature is that plants are sessile organisms, and their growth and development depends on their state and position within the population. Thus the model representation of communities of trees must be seen as a set of discrete individuals located at several points in space. The growth and development of each individual can be described in time as a continuous or discrete function, and interactions amongst them represented as the influence of their nearest neighbours, in terms of their distance, their size, and other structural parameters.

From a mathematical point of view, description of this model population includes applications of differential equations. Firstly, because of the stationary nature of plants, there is no random mixing (i.e. equal probability of any two particles meeting). This is the basis of the application of differential equations in chemical kinetics, and of their transfer as Lotka-Volterra equations into mathematical ecology. Secondly, the size of a population of plants is finite and usually does not exceed 10^4 to 10^5 individuals. This is not a large enough number for appropriate use of mathematical methods based on the idea of the limit transition and infinite size of population. New methods which can be applied in this situation have not yet been sufficiently developed for description of the dynamics of forest communities. Thus large computer simulation models are one of the main features of ecological modelling at the present time.

Ford and Sorrensen (1992) have presented a theory of competition, in the form of evidence for five axioms, which provides a theoretical background for individual-based plant simulation models. In the following paragraphs, their text is in italics.

1. *Plants modify their environment as they grow and reduce the resources available for growth by other plants. This defines the existence of competition.*

The concept of the phytogenic field and its quantitative expression was proposed by Uranov (1935, 1965). This concept will be discussed in more detail later.

2. *The primary mechanism of succession is spatial interaction.*

With regard to this axiom, there are different points of view in forest science and plant population ecology. Succession is generally viewed as a Markov chain with more or less stable transition coefficients, depending on changing environmental conditions or on the previous history of site management. However, another point of view is the idea of primary (endogenous) succession describing mutual changes of the vegetation-soil system. This idea was first proposed by Clements (1916) and much later was developed by Razumovsky (1981). The difference between these two points of view should be recognised. If we consider the sequential changes of vegetation cover, including changes in the dominant tree species, without at the same time taking into account the turnover of carbon and nitrogen in soil, then it is possible to represent these changes as a Markov chain. For primary succession, the situation becomes more complicated, but the main mechanisms of succession dynamics follow the more deterministic rules of a stable system (Razumovsky, 1981). The probabilistic description of succession transforms into algebraic form (Alexandrov, 1992).

3. *Plant death due to competition is a delayed reaction to the mechanism of reduced growth following resource depletion.*

This is the most difficult problem in forest modelling. The modular structure of plants does not allow us to define strictly the meaning of death of a whole tree or plant. This can be done for the dying off of a separate module, a branch of a tree or a ramet of a grass, but the condition of death of a plant as a whole is unknown. Usually, in simulation models, as mentioned earlier, physiologically based dying off is substituted for by random mortality that depends on tree age. Moreover, for grasses (e.g. in secondary dormancy, Rabotnov, 1950), and for trees in a sub-senile state (Popadyuk et al., 1994), it is well-known that individuals which have been rated as having died, and ceased to play any role in the population, can start a new life as a result of the development of new shoots and even of a secondary crown, from secondary dormant buds. It seems to us that partial die-back and delay in death following resource depletion is more complex than any monotonic function of resource deficit.

4. *Plants respond in plastic ways to environmental change, and this affects not only the result of competition, but its future outcome.*

There are many publications reflecting and extending the "plastic response" first described by Harper (1967). In plant population ecology in Russia, the concept of polyvariant development of plants mentioned in Chapter 1 has been elaborated (Zhukova and Komarov, 1990; Popadyuk et al., 1994) and now allows for the explanation in numerical terms not only of different sizes or developmental stages but also of different pathways of development (Zhukova and Komarov, 1990, 1991).

5. *There are species differences in the competition process*

Not all the parameters of a model reflect the processes of competition (i.e. the interactions). We have already mentioned silvics as a means of describing tree species. The silvics may include the most appropriate parameters for sensitive representation of the competition process. For example, they should define shade-tolerance or shade-intolerance, nitrogen response or drought-tolerance. A fortunate choice of the main descriptive parameters may result in a successful theory. We should remember that co-ordinates and impulses, chosen fortuitously as parameters of particles, gave life to statistical physics.

In this Chapter we shall consider mathematical models, bearing in mind spatially explicit, individual-based plant populations or communities, with local interactions among the individuals as a starting point, and try to link the five axioms to the approaches used in these models.

The main approaches in spatial modelling taking these axioms and comments into account may be represented as:

1. the continuous or discrete method for describing plant growth and development,
2. the method for describing local interactions between nearest neighbours, e.g. a distance, square-based, or statistical method.

The models of competition may be of several types:

1. statistical models of spatial structure and their application to populations of trees or other plants,
2. competition indices and spatial tesselations, and
3. analytical models.

We shall trace the historical development of the modelling approaches.

2.2 The Concept of the Free-Growing Tree

Plant-plant interactions, specifically competition, strongly influence plant community dynamics and ecosystem structure. However, little is known about the ecophysiological processes controlling direct plant interactions. If forest management is to be based on sound ecological principles, developing an understanding of these processes, and of plant interactions in general, is important.

It is clear that we have to elaborate the quantitative description of competition. So-called indices of competition have been proposed by various authors; there are good reviews in Tomppo (1986), Pukkala and Kolstrom (1987), Pukkala (1989a), and Ford and Sorrensen (1992). Not all these indices are good predictors of tree growth. One of the reasons for this is that a tree seems to be a system with memory. The growth of a tree depends on its history, which is recorded in its height, biomass, mass of leaves or needles, etc. Thus the growth increment depends only slightly on current values of these indices, with a coefficient of correlation of about 0.3 to 0.4. Moreover, there are some suggestions why plant-plant competition should decrease at a particular age. This property of the stand will be discussed in Section 2.4.3.

Another way of accounting for resource depletion was proposed by Galitsky and Komarov (1974). This is the so-called "concept of the free-growing tree". It was assumed that the key principles governing growth are determined by the balance between the fluxes of matter through the plant. Suppose that the set of factors influencing growth can be classified into two groups as follows: a group ($\alpha_1, \ldots, \alpha_k$), which depends on the physiological activities of the plant; and a group of environmental variables ($\beta_1, \ldots, \beta_m$), which does not depend on these activities, e.g. temperature, light spectrum, etc. A plant may be classified as free-growing if its growth of biomass through ontogeny is affected only by the factors belonging to the second group ($\beta_1, \ldots, \beta_m$).

Introduction of two age-dependent functions is the most important feature of applications of the free-growing concept to tree stand modelling. These are:

$A_{fi}(T)$ – the quantity of the i-th resource necessary for the free growth of the plant; and,

$dB_f(T)/dt$ – the biomass growth rate of a free-growing tree at age T.

This is a mathematical idealization, because such a tree cannot exist within the stand and a tree growing in the open is not the same as a tree within a stand. Strictly speaking, from a mathematical point of view, the free-growing function defines the maximum individual growth curve for a particular site. By analogy with some problems of physics, for example the physics of solids (Ziman, 1960), it may be said that $A_{fi}(T)$ and $dB_f(T)/dt$ describe self-congruence of the plant community as a global variable reflecting the site conditions for a particular species.

The combined effect of all the factors obeys the Liebig principle (Liebig, 1876) that the growth rate is limited at any one time by the factor in which a small change produces a change in the growth rate. It is assumed that similar modifications of other factors do not influence the growth rate. A formalized version of the Liebig principle was proposed by Poletaev (1975) as the theory of L-systems. There is a fundamental link between the Liebig principle and the concept of the free-growing tree.

One suggestion for the construction of a model of biomass growth is that the biomass B(T) is that part of the total biomass with significant exchange of energy and matter.

Using the results of the simple model of Poletaev (1966) and notions of allometry, we can adopt the form of $A_f(T)$ and $B_f(T)$ as

$$A_f(T) = A_0 th^2(T/A_1),$$
$$B_f(T) = B_0 th^3(T/B_1),$$

where th is the hyperbolic tangent, and A_0, A_1, B_0, $(B_1 = A_1)$ are parameters which can be evaluated utilising existing experimental data.

2.3 A Model of a Tree Stand Based on the Free-Growth Equation and Spatial Tesselations

This model of tree stand dynamics is based on the assumption that it is possible to reduce the problem of spatial displacement reflecting resource distribution, and the problem of description of competitive interactions in the stand, to the problem of distribution of the area available for growth. In the individual-based model discussed by Galitsky and Komarov (1987), the area available for each tree in a stand is estimated by constructing the corresponding Voronoi tesselation (Voronoi, 1908).

The basic idea of Voronoi tesselations is that the bounded plane is occupied by a number of points, which are called germs. These germs all start growing at the same time and at the same speed, and gradually occupy the whole space. Whenever different germs meet, growth stops at the point of contact but continues where the space has not yet been occupied by any other germ. The result is called a Voronoi tessellation, with regular boundaries that are straight lines.

In crystallography there is another well-known random tesselation, the so-called Johnson-Mehl tesselation, which was introduced as a simplified model for crystal growth by Kolmogoroff (1937) and Johnson and Mehl (1939). Johnson-Mehl tesselations are perhaps the most tractable and flexible class of random mosaics suitable for plant population modelling.

The geometric structure of Johnson-Mehl tesselations is more complicated than that of Voronoi tesselations. Johnson-Mehl tessellations are obtained if new germs are added while others are already growing. A less regular tessellation allows modelling of complicated contours and can be used for larger scale representation. A big advantage of the Johnson-Mehl representation is its stability. For example, if germs close to the boundary are suppressed, a tessellation is still obtained. In the planar case, the edges of Johnson-Mehl crystals are hyperbolic arcs and three neighbouring crystals may share one or two vertices, while the edges of Voronoi cells are line segments with three neighbouring cells meeting at one vertice only.

Frost and Thompson (1987) studied various models of planar Johnson-Mehl tesselations by simulation. Mahin et al. (1980) present further simulated results for planar Poisson-Voronoi tesselations and for Johnson-Mehl tesselations generated by a time-homogeneous Poisson birth process; and Moller (1994, 1995) shows how typical Poisson-Voronoi cells and Johnson-Mehl crystals can be simulated for a time-non-homogeneous Poisson birth process. Mejering (1953), Gilbert (1962), Miles (1969, 1972), and Moller (1992, 1995) have studied some statistical properties of Voronoi and Johnson-Mehl tesselations.

Applications of Voronoi tesselations for evaluation of relationships between spacing of individual trees and increment of biomass are well-known (Mead, 1966; Rhynsburger, 1973; Liddle et al., 1982; Mithen et al., 1984). The relationship between the square of the Voronoi cell belonging to an individual tree and its biomass characteristics is sometimes surprisingly good, e.g. for young plantations (Martynov, 1976).

The main functional idea in the model is that the biomass located in its Voronoi cell changes independently according to the main equation for biomass growth, until the plant itself, or any of its neighbours, dies (Miroshnichenko, 1955). Redistribution of the square set free after death then takes place; an operation that can be regarded as competition, in this model. The redistributed square is added to the neighbour cells and the process continues. Some features of this redistribution process are discussed by Galitsky (1982).

Analysis of a model of such a community shows the possibility of non-monotonic dynamics of the biomass of the community and its members (Galitsky, 1979), in conditions of sufficient uniformity and density. The dynamics are related to self-suppression of a near-uniform community at certain periods during its development and, consequently, to rebuilding of its mosaic structure. Even if the community members are identical, the occurrence of a dense distribution of trees according to their available areas leads to death of a large number of the trees in the stand, while the remaining trees may continue to grow. Such a situation, i.e. waves of dying off, may be repetitive.

Two main features of spatial modelling are exemplified by this model:

Firstly, it is difficult to define the conditions of plant death caused by its position among neighbour trees. Usually, as mentioned in Chapter 1, death is introduced as a random process. Two main causes for death may be taken into account in individual-based models. One reason is the dependence of the probability of disease on the current state of the tree, e.g. a weakened tree has a greater probability of death. In this case it is necessary to include new processes into the models: diseases, fire etc. The second cause is insufficient account taken of the main features of the growth of the tree, e.g. the exchanges of energy and substances between different components of the tree, such as the redistribution of assimilates between leaves, branches, stem, and roots. In the model discussed above, it was implied that death of a tree depends on the proportion of living biomass relative to the total biomass.

Secondly, the problem of describing spatial *interference* as redistribution of available soil nutrients among plants is highly important but the solution is far from clear. We described the possible distribution of resources as proportional to the distribution of cells in random tesselations, linked with displacement of the points that mark the positions of the trees. The problem of division of continuous resource amongst discrete spatially distributed objects represent a complicated problem depending on the fortunate choice of the method. Another approach, "*ecological field theory*", was proposed by Wu et al. (1985). The main concept in this theory is interference, which is defined as the influence of a plant on its neighbours' environment, through either resource competition or less direct interactions. We shall discuss this approach in more detail in Section 2.6.

2.4 Statistical Analysis of Spatial Structures and its Application to Populations of Trees and Other Plants

2.4.1 Aims of analysing the spatial pattern of plant communities

Spatial pattern in plant communities has been of interest to forest scientists for many years, because it can provide information about stand history, population dynamics, and competition. None of the examples described below constitutes a complete theory, together with significant results, but each serves to demonstrate the possibilities. These approaches show the potential for a complete understanding of the processes determining the dynamics of tree stands or other plant communities. The most interesting problems are concerned with the models for description of interactions, together with the processes of random (or non-random) death. It is necessary that models are developing in which interactive suppression is defined by biomass, height etc. The first attempts to do this are applications of marked Markov spatial processes. The role of analysing of spatial structure of the trees or plants population is to describe the result of interpopulation interactions in general terms reflecting the properties of the spatial structure and its changes.

Another important problem is that development of simulation models, including spatially distributed individual-based models, requires application of Monte-Carlo simulation methods, for which it is necessary to generate the initial tree displacement with values of biomass, height, diameter etc. This problem has been partially solved and will be briefly described in the next section. However, it is linked with the problems of: (i) analysing the dynamics of the spatial structure of stands of different species, taking into account the site conditions; and, (ii) of formulating methods for comparison of the output of simulation models with experimental data.

Spatial pattern analysis is becoming more popular among foresters, and there are many possible applications of these methods in:

- forest inventory,
- forest management planning,
- forecasting forest productivity,
- optimising field measurements,
- diagnosis of the horizontal structure of forest stands,
- finding correlations between tree species and soil properties, and other environmental variables,
- investigation of particular interrelations between neighbouring trees, and effects of competition on spatial structure of forest stands,
- identification of trees, species and other site characteristics from aerial photographs or satellite images, and
- applications to biodiversity and biogeographical problems.

2.4.2 Traditional methods of analysing of patterns of plants displacement

Simple statistical procedures have long been used to address the problem of whether an analysed distribution is random or non-random (Greig-Smith, 1967; Vasilevich, 1969). Non-random patterns may be either clumped (aggregated) or regular (dispersed). There are several different possible patterns of distribution arising from simple statistical models of the displacement of points (Ashby, 1936, 1948; Cole, 1946; Neyman, 1939; Thomas, 1949). In Russia, the first investigation of this problem was by Leskov (1927). There have been many later papers, which were reviewed by Karmanova et al. (1992). Use of these methods is restricted by the complicated procedures of parameter evaluation and the small amounts of experimental data.

The next approach is based on elaboration of some indices, such as the distance from each plant to its nearest neighbour (see e.g. Besag and Gleaves, 1973; Hopkins and Scellam, 1954; Hines and O'Hara Hines, 1965). Surveys of these indices, with Monte-Carlo simulation and evaluation of robustness, were published by Komarov and Grabarnik, (1980) and Karmanova et al., (1992). The indices allow characterization of deviations from randomness of the pattern of points, but they can often not distinguish random and non-random patterns; for example some indices have the same value for random and strongly non-random patterns of points (see Gussakov and Fradkin, 1990, 1992).

2.4.3 Statistical methods based on Markov random fields (i.e. systems with local interactions)

A new stage in the development of analysis of spatial patterns began in the 1970s. New statistical methods were adopted for more detailed investigation of the properties of randomly distributed point processes. One of these was based on the so-called Gibbs random fields and showed very promising similarities with applications in molecular and solid physics. The approach was based on lattice displacement of particles of two different types; interactions between them may be interpreted as attractive (more clumped than random) or repellent (regular), by analogy with ferromagnetic theories.

A new method of evaluating the parameters utilising some appropriate statistical models was proposed by Besag (1974, 1975). The main idea is that the statistical model is built in such a way that the value of a random variable at any point depends on the values at some neighbouring points. For example, let a set of random variables be located at the sites of a two-dimensional square lattice. Then, if conditional probability

$$P(x_{ij} \mid \text{all other values}) = P(x_{ij} \mid x_{i-1,j}, x_{i+1,j}, x_{i,j-1}, x_{i,j+1}),$$

we have a scheme of first-order nearest neighbours for a finite square lattice. The form of P depends on the type of random variable.

2.4.3.1 A binary variable on the lattice

Let a binary (0 or 1) random variable be defined at the lattice sites as an indicator of presence or absence of a plant at the site, then

$$P(x_{ij} \mid \text{all other values}) = \exp\{(\alpha + \beta y_{ij})x_{ij}\}/\{1 + \exp(\alpha + \beta y_{ij})\},$$

where α and β are parameters, $y = x_{i-1,j} + x_{i+1,j} + x_{i,j-1} + x_{i,j+1}$. The estimates of a and b may be evaluated by the maximum likelihood method, on the basis of a single solution for X. This model is a so-called autologistic scheme and is a strong analogue of the Ising model of ferromagnetics (Ziman, 1960), which is well-known in statistical physics. This analogy promotes correspondence between the Gibbs distributions and Markov random fields with local interactions. The parameter a expresses characteristics that are similar to the external field, and it may be easily seen that if $\beta = 0$ then $\alpha = \ln(N_1/N_0)$, where N_1 is the number of trees located at the lattice sites and N_0 is the number of empty lattice sites. The parameter β may be interpreted as a measure of the interaction. In the applications to plant communities, the negative values of β (corresponding to repulsion in the Ising model) may be interpreted as a result of competition which leads to decrease in the probability of survival. The positive values of β may be interpreted as the result of an attraction type of interaction.

The parameters of this autologistic model were evaluated by a unique set of experimental data from a long-term study of forest sample plots (Komarov, 1979), which were set up by Prof. M.K. Turski in 1893, in a Scots pine forest near Moscow (Forest Dacha of K. Timiryazev Agricultural Academy, Eitingen, 1962), when the trees were 28 years old. All the trees located on a regular lattice were enumerated and mapped, and their diameters were measured. The diameters were measured regularly, and dead trees were recorded, during the 55 years from 1893 till 1948.

The results are shown in Figure 4. Parameter a decreased, and β was always less then 0 and varied slightly. In this case, β shows the presence of interactions leading to decrease in the probability of survival, as mentioned earlier. An interesting feature is that the values of β are statistically indistinguishable, i.e. the degree and significance of interactions were constant in time (from the point of view of the Markov random fields model).

2.4.3.2 Markov point processes

In a study by Grabarnik and Komarov (1981), a different Markov model was applied to the same mapped data on self-thinning in the Scots pine stand. In this model the primary population of the site proceeds according to the Poisson process with intensity l, and seedlings arise as a consequence. The probability of survival of a seedling is proportional to $c^{T(r)}$, where T(r) is the number of neighbours located at a distance not exceeding r, and c ($0 \leq c \leq 1$) is a parameter. The trees may die at a constant rate. Let n points $(x_1, \ldots, x_n)$ be located in a bounded area B. Then the density of probability for the set of points $(x_1, \ldots, x_n)$ is $f(x_1, \ldots, x_n)$, proportional to $c^{T(r)}$. It is apparent that the value of c reflects the degree of interaction between the points and log c is similar to coefficient b in the previous model. However, this model allows evaluation of the degree of interaction between points not only for the lattice but also for arbitrarily displaced points. Strauss (1975) proposed a method for evaluation of parameter v (= log c) using experimental data.

The use of two independent models applied to the same data set

Figure 4. Values of coefficient of autologistic scheme for Scots pine lattice sample plot for different years.

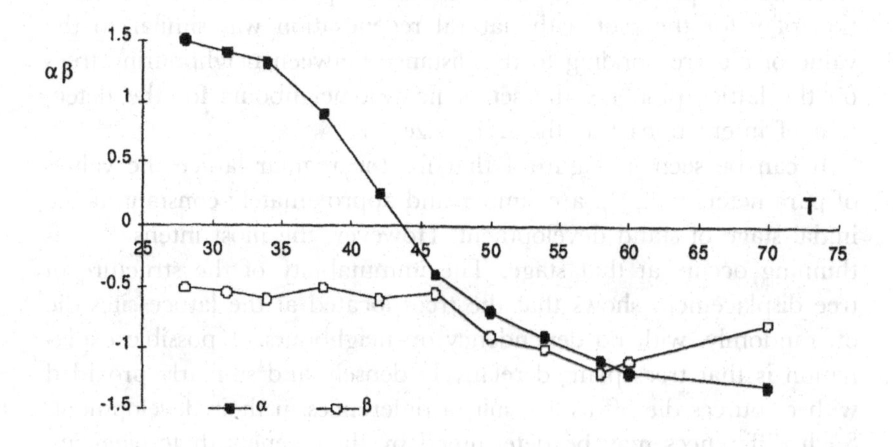

Figure 5. Changes of parameters mentioned in the text for Scots pine plots with natural regeneration (a) and lattice planting (b).

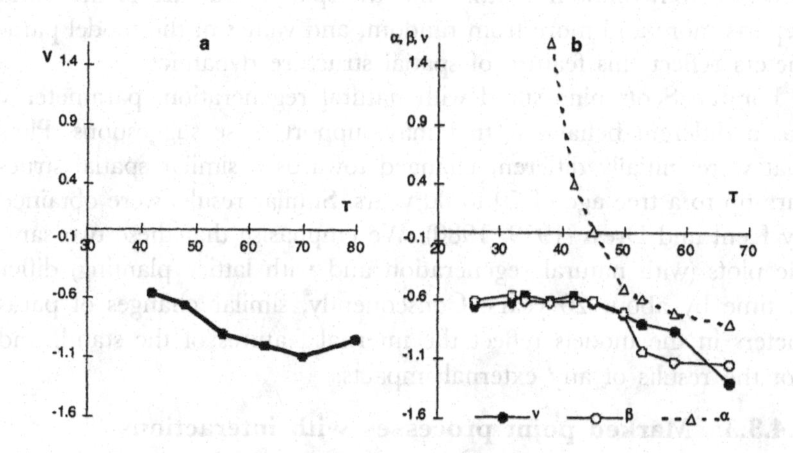

with parameters interpreted in a similar way allows the models to be compared. Data were obtained from naturally regenerated sample plots mapped repeatedly for 80 years and showing the dynamics of intensity of competition mentioned earlier (Figure 5). The same model was applied to the sample plot with trees planted on a regular grid and mapped in the same way as in the previous case.

The changes of the parameters in the models with time are not the result of a statistically dynamic model, but are a successive time series of independently (in time) evaluated parameters. The evaluation of v for the plot with natural regeneration was similar to the value of r corresponding to the distance between neighbouring trees on the lattice plot, i.e. the set of nearest neighbours for the detection of interactions was the same size.

It can be seen in Figure 4 that for the regular lattice the values of parameters v and β are similar and approximately constant at the initial stage of stand development. However, the most intensive self-thinning occurs at that stage. The immutability of the structure of tree displacement shows that the trees located at the lattice sites die off randomly, with no dependency on neighbours. A possible explanation is that trees planted relatively densely and similarly provided with resources die off as a result of differences in their development. Such differences may be determined by their genetic heterogeneity, and not by their position in the lattice. Later, after more have died, heterogeneity of the resources increases and the pattern of displacement begins to determine the inter-tree interactions. Mortality of the genetically selected trees is determined by the spatial structure and deviates from random death. Thus the spatial structure of the stand departs more and more from random, and values of the model parameters reflect this feature of spatial structure dynamics.

For the Scots pine stand with natural regeneration, parameter v has a different behaviour that may support these suggestions. Plots that were initially different changed towards a similar spatial structure up to a tree age of 70 to 80 years. Similar results were obtained by Kent and Dress (1979, 1980). We emphasise that these two sample plots (with natural regeneration and with lattice planting) differ in time by about 20 years. Consequently, similar changes of parameters in the models reflect the internal features of the stand, and not the results of any external impacts.

2.4.3.3 Marked point processes with interactions

Further extension of estimation methods based on Gibbs (= spatial Markov) models has been related to evaluating interactions as a function of distance between trees. Generalisation of the models to marked Gibbs models also results in a remarkable gain in flexibility, and better use of the experimental data (Goulard et al., 1996).

A more complex statistical model, taking into account the Gauss random variable, linked with spatially distributed points, has also been proposed (Besag, 1974). This model differs from the binary random variable model marking presence or absence only and provides an opportunity to find some interaction parameters through tree characteristics such as height, biomass and crown size.

In this model x is a Gauss random variable located on the bounded regular lattice at the site (i,j). Then mean

$$\mu' = \mu + \beta_1 (x_{i-1,j} + x_{i+1,j} - 2\mu) + \beta_2 (x_{i,j-1} + x_{i,j+1} - 2\mu)$$

where μ is a constant coefficient, and mean x of all values, and the variance, is also likely to be constant. This scheme has been named autonormal (Besag, 1974). Parameters β_1 and β_2 can be interpreted as a measure of interactions with nearest neighbours. The difference between β_1 and β_2 corresponds to the difference between two perpendicular directions at the lattice. If $\beta_1 = \beta_2 = 0$, then there are no interactions between lattice sites. We have independent solutions for the Gauss random variable x. If at least one value of β_i is not equal to 0, then there is some inter-dependence between nearest neighbours. If β_i is negative, then interactions may be defined as competition, since, in this case, the larger the nearest tree to the site (i,j) the smaller the mean for the site. The computing procedure is complex. Some attempts have been made to process experimental data (Komarov, 1979).

Data on three forest species (even-aged oak, birch and pine) planted in a regular lattice (in Pushchino city park) were processed using this autonormal scheme and some clear results were obtained (Table 3). The intensity of competition was different in these plantations and a good picture of morphological differences and other signs of competition was obtained. Unfortunately, this model has limitations with respect to this application because of the special form of the experimental data.

Table 3. Parameters of interactions between nearest neighbours for different species, evaluated by the Markov autonormal model.

Species	M × N	β_1	σ_1	β_2	σ_2
oak	42 × 7	-0.2339	0.053	-0.2441	0.084
birch	24 × 10	-0.1643	0.025	-0.3139	0.030
pine	13 × 26	-0.2066	0.024	-0.2730	0.011

β_1 and β_2 show the influence of neighbours in the directions west-east and north-south, respectively. For oak, $\beta_1 = \beta_2$, i.e. interaction exists but does not depend on the light conditions. This is as expected, because these oaks were about 30 years old, at a spacing of 5 metres (the interval of the lattice), and the crowns were not interacting. In contrast, for birches interaction is evident as different values of the coefficients. The birches had an average height of about 15 metres, so that shading of one by another was intense. For pines, the situation was intermediate.

Applications of these lattice statistical models with local interactions may be extended to situations with arbitrary displacement of trees (Besag, 1975). There are too few publications about applications of Markov field models, in spite of their potential. The main reason for this may lie in difficulties with the calculation procedures and the complicated mathematical form of their presentation, mostly published in mathematical journals.

2.4.4 Analysis of spatial pattern using second order statistics

Another method of analysing a spatial structure was proposed by Ripley (1976). He introduced the special K-function which uses not only the so-called first order statistic, i.e. the mean of all plant-to-plant distances, as in nearest-neighbour methods, but also the variance of the distances, the second order statistic. In spatial pattern analysis, a circle of radius t is centred at each point and the number of neighbours within the circle is counted. For n individual points distributed in an area A, the density ($\lambda = n/A$) gives the mean number of plants per unit area. The function $\lambda K(t)$ gives the expected number of further points within radius t of an arbitrary point. If the points are randomly (Poisson) distributed, the expected value of K(t) equals πt^2, i.e. the area of a circle of radius t, and a plot of $\sqrt{K(t)}$ against t should therefore be linear. K(t) was tested against the null hypothesis of complete spatial randomness (Diggle, 1983), in which all points are assumed to be distributed independently. Other hypothetical models can also be tested (Ripley, 1979; Sterner et al., 1986; Kenkel, 1988). A brief survey, together with examples of applications of Ripley's K-function, was published by Haase (1995).

This method is well-known and has been applied in Russian forest science. The method proposed by Busykin et al. (1985, 1987) is similar to Ripley's K-function. The K-function was applied to analysis of the spatial structure of tree stands by Grabarnik et al. (1992), who described an extension of the K-function and applied it to processing experimental data on Scots pine and Norway spruce stands in Central Russia. The results show the possibility of detecting stability or disintegration of a stand.

Another application of this tool for analysis of a tree distribution pattern and its connection with tree characteristics was that by Gavrikov et al. (1993), who analysed sensitivity of such characteristics as tree diameter and height with respect to spatial displacement of the nearest trees. They showed that neighbouring trees affect asymmetry of the centre of the crown of a tree with respect to the trunk in Siberian pine forest and suggested a model for describing this effect.

2.5 Modelling of Spatial Structure

Analysis of spatial patterns of forest stands gives rise to the particular problem of modelling a stand with known spatial structure and tree size characteristics that depend on local conditions. This is of great importance because of the need to link experimental data with simulation models at population level, mainly individual-based models. It is well-known that spatial displacement of trees in the stand defines the interactions amongst the neighbours. As we mentioned earlier the dynamics of stand starting from regular displacement in a case of forest plantation differs from the stand generated by natural regeneration (Kent and Dress, 1979, 1980). In the case of Monte-Carlo simulations of the spatially-explicit individual-based models we must simulate set of initial patterns with the same properties of spatial structure and linked size characteristics.

With the models we have described earlier in mind, we see that this problem is more complex than merely characterizing spatial structure from the point of view of any existing statistical model. Some attempts were made in the 1960s and 1970s (Newnham, 1968; Oikawa and Saeki, 1972), based on distance indices. However, these indices often prove to be inconsistent.

The description of the spatial structure of a tree stand or other plant community is a necessary component in an individual-based model of the community. We introduce spatial structure as a set of data about N trees (or other plants) in an area W, with co-ordinates (x_i, y_i) and vectors $\{Z_i\} = (z_{i1}, \ldots, z_{ih})$ attached to these points as a set of values of characteristics used in the model. This usually corresponds to an experimental or simulated sample plot.

Simulation models which describe spatial structure in this way may be classified as:

- models of competition, stand growth, and forest succession;
- models of radiation regime in non-homogeneous vegetation;
- models of genetic structure of tree populations (Galitsky et al., 1982); and
- models of spatial structure of stands for assessment of the representativeness of forest inventory data.

An attempt to build a general theory of computer experimentation with generated spatial patterns was made in the recent book by Gussakov and Fradkin (1990). We describe briefly their main ideas about simulation of spatial structure with applications to plant communities.

If Z_i represents a random vector, then the spatial pattern corresponds to a random field defined by the set of points. Such structures are called marked point processes (Stoyan et al., 1987). Let $\psi = \{[p_i, Z_i]\}$, where p_i is a point on the plane and Z_i is the mark (a vector in the general case) of p_i, ψ_w is the realization of the marked point process corresponding to sample area W. The simulation model may be seen as the operator L defining the correspondence between ψ, describing the spatial structure of the marked point process, and an output of the system Y in the form of a random variable, a random process, or a random field.

For instance, the phase trajectory of some model of stand dynamics, taking into account the spatial pattern of trees with a fixed scenario of external impacts, may depend on the particular representation of the spatial structure of the stand, and thus is a random process. Because of the random behaviour of the output of simulation models for different spatial patterns, the problem of analyzing the dependence of this output on the statistical characteristics of the marked point process ψ and the operator L must be solved by Monte-Carlo methods. This is often not appreciated, so that dependence on the different spatial solutions to the Monte-Carlo process is not analyzed.

The general algorithm for the generation of spatial patterns, which are specified by parameters describing the attributes of the pattern, may be formulated in the following way (Gussakov and Fradkin,

1990). Let some criterion A characterizes the pattern attributes with value A_0. The point pattern is required such that $|A' - A_0| < \varepsilon$, where A' is the current value of criterion A for the simulated pattern, and e is the accuracy of the calculations. In the first stage of simulation, N independent pseudo-random values with uniform distribution are generated at (0,a). Each point may attract or repel the points which occur inside the circle of radius R, with the centre at p_i. The points outside the circle are not influenced by p_i.

We can define the function of attraction-repulsion, f, in such a way that $f > 0$ corresponds to repulsion, and $f < 0$ to attraction. Let $| f | p_i$ = the distance of shifting p_j as a result of the influence of p_i, and r_{ij} = the Euclidean distance between p_i and p_j. If a regular pattern forms, then $f > 0$, and $f = 0$ if $r_{ij} = R$. The simplest form of f which satisfies these conditions is $f = R\text{-}r_{ij}$, then p_i tends to push point p_j to the boundary of the circle. For the generation of clumped patterns, we must define the function which describes the attraction of p_j to p_i, so $f = -r_{ij}$.

Uniting these two functions and introducing the additional formal parameter γ, we get

$$f = (1\text{-}\gamma)R\text{-}r_{ij}.$$

Then at $\gamma = 0$ we have repulsion (regular pattern), and at $\gamma = 1$ we have attraction (clumped pattern).

Gussakov and Fradkin (1990) proposed the method of changing the initially random positions of points in relation to a chosen value of γ. The main algorithm for simulating patterns with new properties consists of the following procedures: 1) choice of criterion and its value, A_0, 2) setting the accuracy of approximation ε, 3) setting the key parameter γ, which defines the type of pattern (regular, random, clumped), 4) setting the limiting radius of interaction, R, 5) choice of Δ, a time-step similar to that in numerical methods of solving differential equations.

The chosen criterion A has to possess good statistical quality; primarily it has to be robust. It was shown by Grabarnik and Komarov (1980) and by Gussakov and Fradkin (1990) that the Hopkins test is the most robust of the criteria using distances between points.

Other methods have been proposed for generating patterns with known statistical properties. The best-known is the method of "simulated annealing", which is based on thermodynamic concepts (e.g. Kirkpatrick et al., 1983).

This method can be applied to simulated patterns of plants with known K-function. The problem can then be formulated as the development of a computing procedure using a known K-function $g^0(r)$ that can simulate patterns with K-function $g^0(r)$ that satisfies the condition $A^* = g^0(r) - g(r) \mid < \varepsilon$, with constant average number of points per unit area.

These methods of simulation of spatial patterns with linked individual-tree characteristics have a mathematical background which is largely based on the theory of Markov random fields. Many other methods based on other ideas have been proposed for simulating spatial patterns. An interesting method for generating a range of two-dimensional point patterns, from regular through Poisson to clumped, was proposed by Wu et al. (1987). Similar methods were described by Cox and Isham (1980) and Diggle (1983). Interesting methods based on competition indices for predicting tree dimensions, in such a way that diameter, height and age are properly related to each other and to the spatial distribution of trees, were examined in a series of papers (Pukkala, 1988, 1989a,b; Pukkala and Kolström, 1987, 1991; Miina et al., 1991; Pukkala et al., 1994), in which possibilities for the application of competition indices were comprehensively discussed.

2.6 Spatial Analytical Models of Competition

A relatively large number of models of vegetation dynamics have been developed, e.g. neighbourhood patch models (Weiner, 1982, 1984), models of plant population dynamics (Pacala and Silander 1985, 1987; Silander and Pacala, 1985, 1990), and others. Some of these models were designed to elucidate the basic mechanisms underlying vegetation dynamics. In simulation models, such as JABOWA (Botkin et al., 1972; Botkin, 1981, 1993), FOREST (Ek and Monserud, 1974), and FORET (Shugart and West, 1977), plant-plant interactions are greatly simplified in order to provide more detail about long-term community dynamics. JABOWA, FORET, and other gap models (Shugart, 1984) are not spatially explicit at the scale of an individual tree, so that there is no means of representing spatial heterogeneity in soil resources and the relative locations of the simulated plants. Most gap models also treat individuals as the average of the population and do not take into account differences in size and proximity within patches.

We restrict ourselves by reviewing another class of population models in which the population is treated as a set of spatially displaced sessile objects with local interactions that is we shell discuss the models dealing the description of competition among plants.

A definition of competition still used today is that given by Clements et al. (1929, cited by Ford and Sorrensen, 1992): "When the immediate supply of a single necessary factor [for growth] falls below the combined demands of the plants, competition begins". There are a large number of different heuristic quantitative descriptions of competition between trees and other plants. Different classifications, highlighting different approaches, have been developed. There is a detailed and constructive review by Ford and Sorrensen (1992), and comprehensive reviews by Benjamin and Hardwick, (1986) and Benjamin and Sutherland (1992), in which many models developed to describe the growth of individual plants in crops and stands have been classified according to the assumptions made about how resources are shared.

There are interactions among individual plants located within a patch (Weiner, 1990), and these interactions are indirectly influenced by the proximity and size of neighbours (Harper, 1977; Solbrig, 1981; Silander and Pacala, 1985; Weiner, 1982). Consequently, the mechanisms of plant-plant interactions are obscured by ignoring such local interactions as a function of the position of individual plants within a community and of the size of the community (Huston et al., 1988; Silander and Pacala, 1985). Furthermore, the competitive response of a plant is not directly related to its neighbours, but to resource availability (light, soil moisture, and nutrients), i.e. it is determined by the abiotic environment and modified by the structure of the patch (Bazzaz and Sipe, 1987; Goldberg, 1987, 1990). Thus mechanistic models of competitive interactions should be individually based, spatially explicit, and resource-driven (Mou et al., 1993).

Despite the lack of a general analytical theory of competition, some models that explain empirical observations have been developed in the last two decades. Here we include as analytical those models that describe population dynamics on the basis of suggestions about the nature of the main mechanisms. We restrict ourselves to examination of individual-based models that describe interactions defined by mutual displacement in a patch. These models can be divided into three groups:

- models of dynamics of the tree biomass characteristics;
- models describing mutual growth of mixed populations of two species; and
- ecological field theory models.

2.6.1 Spatial Models of Tree Biomass Characteristics Dynamics

Experimental data show that distribution of tree mass, tree diameter, etc. may be bimodal (Ford, 1975). Moreover, the dynamics of the shape of distributions is based on simple biological mechanisms (Komarov, 1979; Huston and DeAngelis, 1987; Ford and Sorrensen, 1992).

In the initial stage of stand development, diameters may be described by an exponential distribution. Then, as a result of the growth of the tree and of self-thinning, the distribution changes to an asymmetrical, unimodal distribution similar to a logarithmic Gaussian or Railegh distribution. The trees with crowns in the upper canopy are then growing faster than the suppressed trees in the lower canopy, and the distribution becomes bimodal. The spatial pattern of the live trees changes from Poisson towards the more complex structure briefly discussed earlier.

Ford (1975) initiated the mathematical description of this phenomenon, which was clearly the consequence of spatial interactions among nearest neighbours. Diggle (1976) developed a one-sided interaction in which the influence of a large plant was dependent upon its size and the distance to its neighbours. Competition on a triangular lattice was considered between individuals and their nearest six neighbours, each as pair-wise encounters. Some simple suggestions were made which allow the construction of a mathematical model able to describe these transformations of biomass distributions.

These suggestions are: 1) a zone of influence for each tree corresponding to a circle with radius proportional to the diameter of the tree; 2) interaction amongst individuals defined by overlapping of zones of influence, with the degree of interaction depending linearly on the distance between the target plant and its neighbour in a way that specifies the asymmetry of large plants influencing small ones, but with no reciprocal effect; 3) definition of a set of nearest neighbours, and for each plant a competitive status that is multiplicative with respect to the surrounding neighbours; and 4) any two plants considered to be located at equal distances which are the lattice steps.

The effect of asymmetry was marked. As distance between neighbours decreased, with other parameters constant so that competitive intensity became stronger, distributions progressed from unimodal (no competition), to bimodal, with a relatively large number of small plants. With further decrease in distance, distributions became bimodal, with a larger proportion of large plants, and then, for very small distances, unimodal again. These changes of size distribution are not dependent on time but on distances between neighbouring plants, so that this is not a dynamic model.

A dynamic model was developed by Gates (Gates, 1978; Gates and Westcott, 1978). In this model, plants are considered to be growing at the sites of a regular lattice and competition is defined as overlapping zones of influence, with a non-overlapping zone defined as a zone of survival. If the zone of survival becomes equal to zero, the plant dies. The model allows analytical tractability and includes time dependence. An initially unimodal distribution gradually becomes bimodal. Bimodality of the resulting size frequency distributions is predicted if, and only if, the area of overlap is assigned to the larger of the two competing plants. Gates (1978) found parameters for the growth rates of plants that give developments in bimodality over time which are similar to the experimental data of Ford (1975).

There are some developments of lattice models that allow relatively rigorous, formalized mathematical constructions, e.g. Gates et al. (1979), Gates (1982), McMurtrie (1981), Ford and Diggle (1981) and Cannell et al. (1984). These studies provide explanations for some of the imperfections of competition indices, and of the important role of a three-dimensional approach for description of competition. (See the review by Ford and Sorrensen, 1992).

2.6.2 Models of two competing plant species

The second group of analytical models attempts to define the relationships between two competing species growing at the same site. Similar problems in animal ecology have resulted in the famous predator-prey equations by Lotka (1925) and Volterra (1931). There are no similar equations in plant ecology but there have been many attempts to solve the problem of coexistence of two or more plant species. The common feature of these attempts, and the way in which they differ from the Lotka-Volterra equations, is that equations only for stationary conditions are relevant to plants. Thus attempts to construct dynamic (time-dependent) equations fail. It is important to keep this in mind, because the plant community is more complex than the animal community.

The best-known equations for two competing species are the de Wit equations (de Wit, 1960). In this model, plants of the two species are located at the vertices of a square lattice.

Here, the main mechanism of competition is occupation of the square of nutrition. It is possible to derive from the model some important characteristics of the competition process, such as the coefficients of exclusion, showing the ratio between the necessary squares of nutrition of each species. The productive capacity, i.e. the ratio of yield to seed sown, is also an important characteristic. The species which is able to occupy the largest square, or gives the highest yield in a smaller square, is successful in competition. Although this approach is largely empirical, with a large number of phenomenological coefficients, k_{ij}, De Wit's model describes experimental data well.

Gates (1980a,b) generalized the de Wit model. He showed that the phenomenological coefficients can be calculated on the basis of simple suggestions about the mechanisms of competition between plants. The most valuable aspect of Gates' model is that it was constructed to examine the behaviour of a large statistical assemblage of homogeneous, similar, interacting plants. By examining the properties of such an assemblage, Gates depicted the phenomenological nature of de Wit's equations and this is very important because of the clear link between de Wit's phenomenological approach and Gates' rigorous statistical results.

Fisher and Miles (1973) analyzed crop plants growing at the sites of a regular lattice and weeds located randomly among these sites. They also introduced the concept of a zone of nutrition, which is a circle which expands at a constant rate. They examined, firstly, the shape of the boundaries between zones, and, secondly, the influence of lattice structure, various rates of expansion, and the times at which weeds arise. They showed that: a) there are optimal interrelations for plants amongst all the parameters mentioned above; b) the form of the lattice is very important, and the optimal form is a square lattice, because for lattices which are not square the weeds are successful; and c) the rates of expansion of zones of nutrition are also important. It is significant that their results are in agreement with those of de Wit discussed above.

The equation for two species growing together was first suggested by Uranov (1935, 1955). Consider a two-species plant community, in which the first species, the active species, influences the second, passive, species. If we have a set of sample plots of equal size, then we can sort these plots on the basis of the abundance of the active species. Using many experimental data, Uranov showed that the correspondence between the two species can be expressed as

$$y = y_0 x^b e^{c(x-1)},$$

where x and y are abundances of the active and the passive species, respectively, and y_0, b, and c are parameters evaluated from experimental data. All the coefficients in this equation have a definable biological meaning. Uranov called this the contingency equation and described five types of contingency: 1), positive (if x increases, then y also increases), 2), negative (if x increases, then y decreases), 3), double-signed (if x increases, then y first increases, and then decreases), 4), complex (if x increases, then y first decreases, and then increases), and 5), indifferent (changes of x do not result in changes of y). All these types of contingency may be approximated by the contingency equation with appropriate values of the coefficients.

Using slightly different suggestions about interspecific interactions, Uranov (1965) modified the contingency equation into

$$y = y_0 e^x e^{ax-c},$$

This modified contingency equation also describes all the types of contingency. Numerous examples of particular forms of contingency have been developed (Serebryakova, 1977).

Uranov introduced the concept of the phytogenic field, defining it as the part of space where the environment acquires new properties as a result of the existence of a particular plant in the area (1965, p. 251). He proposed to measure the phytogenic field of a particular plant by taking into account the frequency of encounters with plants of another species in relation to the distance between them. Different forms of this relationship have been investigated in natural communities (Uranov and Mikhailova, 1974; Zaugolnova et al., 1988). It is interesting to note that the concept of the phytogenic field is very close to the idea of Markov random fields, particularly in the methods of potential description of local interactions with nearest neighbours. This similarity may be very useful for finding the general form of competitive interactions.

2.6.3 The ecological field theory

The idea of using classical field theory from physics has been proposed as the basis of "the ecological field theory" (Wu et al., 1985, 1994; Sharpe et al., 1985; Olson et al., 1985; Walker et al., 1989). The main concepts on which this theory is based include interference and interference potential. Interference is the influence of a plant on the environment of its neighbours through resource competition or less direct interaction.

The different effects of canopy, stem, and roots on site resource availability are calculated separately and then summed to give an estimate of total plant effect. In non-interacting situations (i.e. without overlapping planar domains), the resources available to a plant from a given location within its domain depend on the resources that can be supplied. However, in interacting situations, if the resource availability from that location is not sufficient to meet fully the requirements of interacting plants, then resources are partitioned to interacting individuals, based upon the intensity of their influence. The final soil resource availability (i.e. water or nutrients) for each individual is calculated by summing and normalizing the available resources to that plant at all locations within its domain. Light availability for any individual plant is determined on the basis of its height and leaf distribution, and the height of interacting neighbours. The resources (light, water, and nutrients) acquired by a plant are then introduced into a continuous-time Markov equation to calculate growth. Thus this approach treats competitive effects and responses separately, with each interacting closely with the other through time.

This concept may be applicable to ecosystems with the following characteristics: a), each organism or group of organisms competing for resources has a specific location that can be identified by a set of (x,y) co-ordinates; b), the influence at a distance is identifiable and the intensity of this influence with distance can be described mathematically; c), where more than one resource is identified as important, the influence of a resource must be able to be scaled for coupling with other resources; and d), an appropriate quasi-steady state life-process model can be derived.

Wu et al. (1985) developed six equations that serve the following functions in the model: 1), to determine the spatial effect of an individual plant on each particular resource (light, water, and nutrients); unique effects attributable to roots, stem, and canopy are calculated separately; and 2), to calculate the residual resource availability after plant uptake at each site. Other equations developed by Mou et al. (1993) calculate resource availability (light, water and nutrients) for each individual plant at each simulation time step. Water and nutrient availability are calculated by summing the resource partitioned to the plant, based upon its relative influence at each location within its domain. Light availability to any individual is estimated by summing the products of relative leaf proportion and relative light intensity, as described by Beer's law for each canopy layer.

After resource availability to each individual is determined, the growth rate and changes in dimensions are simulated for each plant. The continuous-time Markov model is a quasi-physiological simulation of the photosynthetic process and is described in detail by Olson et al. (1985), Mou et al. (1993) and Wu et al. (1994). The full system of equations for application of ecological field theory is described in detail by Mou et al. (1993).

Although this theory, and the modelling approach, are data-intensive, they provide general guidance for studying plant-plant interactions more mechanistically.

3 Analytical Models of Tree Populations and Forest Communities

The purpose of this chapter is to give a brief survey of a newly formulated theory of mathematical modelling of forest community dynamics, based on the following components: a) the layer-mosaic concept of spatial-age structures of forest communities, the so-called gap paradigm; b) computer simulated gap-models; and c) the theory of structural models of populations and communities. This new theory allows investigation of the dynamics of forest communities at various spatial and temporal scales.

3.1 The Hierarchical System of Mathematical Models of Forest Communities

The forest community is very complex. According to the general principles of system analysis, it is necessary to construct not one model, but a hierarchical set of models for the mathematical description of the structure and dynamics of these complex systems, such that a model at a particular scale includes models of the previous scale as elementary objects or ready-made modules. The principles of construction of such a hierarchical system can be different, for example, with respect to spatial scale, significant time-periods of the dynamics, or levels of structural organisation.

We may consider a number of mathematical models of forest communities as components of the hierarchical system of models of tree populations and communities, constructed on the basis of increasing complexity of structural organisation and corresponding expansion of spatial and temporal scales, namely: a single tree – locus (or gap) – population – meta-population – community or forest landscape. We will now briefly consider the basic stages of this system; a full description will be given later.

1) The initial item in this system is a model of a free-growing tree, such as the models of Poletaev (1966) and their updated forms (Karev, 1984), the model of Galitsky and Komarov (1987), or any model of tree growth used in gap-models (e.g. JABOWA, FORSKA etc.).

2) The transition up to sub-population scale (locus, gap) is made by synthesis of the free-growing tree model together with models of competition for external resources, such as sub-models of SOM dynamics and competition for light. A model of the dynamics of a single generation of a tree sub-population was the result of the synthesis e.g. the various separate gap models (Shugart, 1984). Another example, dynamic models of tree growth, DMTG, (Karev and Skomorovsky, 1997a), is briefly described below.

3) Accounting for the processes of interactions amongst the trees is the basic difficulty with construction of local dynamics models. A choice of initial distribution, or making some assumptions about the current distribution of the trees, allows this difficulty to be overcome for models of single generation sub-populations. The simplest assumption is that all trees in a sub-population are identical (such a model describes the average tree growth). However, models of non-uniform populations are more realistic and for this purpose the following classes are important: a) the initial distribution approximates a step-function (i.e. it corresponds to a model of a multi-layered tree stand); b) the initial distribution is uniform in some areas (this assumption has been made in many computer gap-models); and c) the initial (or current) distribution has negligible small moments of orders more than k (usually $k = 2$). The transition from a model of a tree sub-population, consisting of identical trees, to a model of a non-uniform sub-population, is the first stage of spatial scale expansion of a gap-model.

4) The layer-mosaic concept considers a forest community as a meta-population consisting of a large number of loci or gaps. This approach allows the next step in enlarging the spatial scales of the model to be taken. For realisation of this approach, it is necessary to find rates of birth and disappearance of individual gaps as integrated formations. The completion of this task essentially depends on the particular gap model. It appears that such fundamental characteristics as probability of disappearance, duration of gap life, and also life duration of a whole meta-population, depend on initial gap size and other characteristics of tree distribution (e.g. variance). These results again emphasise the critical importance of gap size, as has been pointed out both in theoretical works and in experiments with computer gap-models (Shugart, 1986; Popadyuk et al., 1994). Furthermore, the various classes of meta-population models may be constructed by the same methods as were used for construction of models of non-uniform sub-populations, if "tree" is replaced by "gap". Models of

non-uniform meta-populations (including spatially non-uniform ones) are the most interesting.

5) At the next stage, it is possible to study model behaviour over long time intervals. It has been shown on the basis of rather general assumptions (Karev, 1992), that (irrespective of the specification of a separate tree model, and independently of the competition and interaction of trees within the sub-populations), a unique stable (climax) distribution exists for the meta-population as a whole. This is then the limit of the process of fast convergence from any initial state. The limiting distribution describes an unique stable state of dynamic balance, representing a mosaic of gaps of various ages and states, which are formed, develop and disappear according to internal laws, but the distribution of all the sets of gaps turns out to be stationary. Obviously, this statement does not contradict results which indicate that an oscillatory mode in a tree population may be established as a consequence of an oscillatory regime within a sub-population that is an element of a wider meta-population.

A research of asymptotic behaviour of large simulating models is not possible without its inclusion in frameworks of the analytical theory. Such inclusion can be done on the basis of the thesis: *Gap – models can be considered as computer realisations of the structural models of meta-populations and communities.* The results about structural models of meta-populations compose a basis of the mathematical theory of gap – modelling. The theory gives the possibility to study directly the asymptotic behaviour of the landscape gap-model under different parameter values, various initial distributions, etc.

6) The top level of structural organisation and spatial-temporal scales includes the models that describe succession processes. There are two basic classes, gap-models and the Markov models of succession (Shugart, 1992). These models are special cases of a new class of structural succession models currently being developed (Karev, 1994, 1996).

The structural succession models allow to investigate the important "ergodic properties" of the climax state of succession association of forest communities, which proved to be connected closely with the fundamental problem of stability and biodiversity of forest ecosystems. According to the "ergodic theory" ideas, the climax state of succession association may be considered as unwrapped in space time history of the association.

A new theoretical method of complex estimation of forest ecosystem states follows from "ergodic properties" (Karev, 1997). It allows to estimate in details a deviation of an observable state of the association of forest communities from the steady climax state.

3.2 Analytical Models and Basic Principles of Their Construction

Modelling of forest stands was originally based upon empirical models describing the changes of stand characteristics (i.e. number of trees, total biomass, etc.) over time. Perhaps the first analytical model of forest vegetation consisted of the simple equations of stand volume increment and energy flows in forest communities published by Khilmi (1957, 1966, 1976). The model was comprehensively analysed by the author in relation to flows and pools of solar energy in the forest stand. Later, the plant community was considered as a system consisting of a large number of interacting individual plants. The models were constructed as some sort of synthesis of individual plant models and models of plant interactions. The model PUU-1 (Kull and Kull, 1989), describing the growth of coniferous trees, is a good example of a detailed ecophysiological model, which should be regarded as a theoretical research tool because of its size and complexity.

Some general principles of construction of an individual plant model have now been formulated.

a) *The construction of mass balance equations* is one of the basic methods of modelling. The balance equations, usually based on conservation laws of mass and energy, are applied in the vast majority of modern plant models, which take into account the division of plants into separate components. Balance equations in dynamic models usually have the form of differential equations connecting the rates of the processes, often according to the following principle.

b) *The principle of limiting factors* states that the rate of a process is determined by the minimal rate of a sub process at any moment in time. According to Liebig (1876), plant growth rate is limited at any instant by those factors in which a small change produces a change in the growth rate. It is assumed that similar changes in other factors do not affect the growth rate. Poletaev (1973) has enlarged this principle to the case in which the limiting component is not only a concentrations but may also be a flow of substance or energy. In such a form, the principle of limiting factors, formalised in the theory of L-systems (Gilderman et al., 1970), has become one of the basic concepts utilised in plant growth modelling. A number of plant models have been constructed on the basis of the systematic use of this principle (Poletaev, 1973, 1975; Kudrina, 1973). Poletaev has attempted to construct a basic physiological plant model describing the whole history of development from seed to seed, but unfortunately this has not been completed. A variant of this model (Treskov, 1987) contains more than 500 equations and is divided into submodels constructed within the framework of the mathematical theory of L-systems on the basis of modern physiological knowledge.

c) The *optimisation principle* may explain adaptive qualities of biological objects. This promising and evolutionarily sound principle has been poorly used, largely because of mathematical difficulties. The optimisation strategy of an organism can be of two kinds: (1) maximisation of productivity, and (2) maximisation of reproduction. The first strategy was used by Rachko (1979), Oja (1985b, 1986), and Kull and Kull (1989) for optimal ontogenesis modelling. They assumed that the distribution of photosynthetic assimilates in a tree is such as to promote maximum biomass increment. The second strategy corresponds to the Holdane-Semevsky principle of the differential survival of an organism (Semevsky and Semenov, 1982) and has been much used (e.g. Korzukhin, 1985; Korzukhin and Ter-Mikaelian, 1987; Korzukhin and Semevsky, 1992).

d) *The law of Zipf – Paretto – Mandelbrot.* Zipf (1949) developed an empirical rule describing range distributions of different kinds. Initially, the rule was discovered in the following situation: the words in a text are ranked in order of decreasing frequency, words with the same frequency being ordered arbitrarily. Let r designate the rank assumed by a word of probability P. Then $P \approx 1/r$, or, more exactly (Mandelbrot, 1977):

$$P = F(r+V)^{-1/D} \qquad (3.2.1)$$

where three parameters D, F, V are related by $F^{-1} = \Sigma\,(r+V)^{-1/D}$. In other words, the graph of (log P vs log r) is a straight line. Later, it was discovered that this law has many applications, from sociology to biology. In particular, it allows prediction of non-observed values, using the distribution of observed values. For example, it can be used to predict root biomass using data on above-ground tree biomass (Sukhovolsky, 1996).

This rule was one of the sources of the development of fractal theory (Mandelbrot, 1977), which has become an intensively used methodological tool for forest modelling. Recently fractal models have been successfully applied in analysis and representation of structure of tree crowns (e.g. Zeide and Pfeifer, 1991; Berezovskaya et al., 1993, 1997; Gurtsev and Tselniker, 1997). A Zipf-Paretto-Mandelbrot law (3.2.1) has now become a general principle in community modelling.

e) The *principle of allometry* (Huxley, 1932) has resulted from attempts to explain the rather stable relationships between size and mass of a tree in the course of ontogenesis. In practice, relationships between individual organs are maintained during their growth and can be expressed as a power dependence. The principle of allometry has been demonstrated by numerous empirical data obtained in studies of dendrometry and productivity and it is now one of the most convenient and widely used principles in both analytical and simulation modelling of forest stands and communities. The relationships allow the possibility of calculating the mass or volume of the organs of a tree using empirical species-specific proportions between the parts, expressed in the form

$$y = ax^b. \tag{3.2.2}$$

Allometric relations enable a considerable reduction in the number of independent variables of a model. However, this approach should be applied with caution, because the relationships may change from one situation to another (Terskov and Terskova, 1980). There have been some interesting attempts to substantiate the principle of allometry theoretically (Kofman, 1981).

These principles have usually been applied in the majority of ecophysiological models commonly used over the last two decades, both in separate tree and in individual-based population models, including simulation gap-models. These models are the best ones for modelling of forest communities in specific external conditions for growth on rather small spatial and temporal scales. However, calculation of the dynamics of each separate gap (and even of each individual tree) turns out to be a major limitation for forest community modelling on regional and landscape scales, and also over long periods of time. Ways of overcoming these difficulties in the theory and application of gap-models were discussed by Karev (1994) and will now be considered further.

3.3 Example: Dynamic Models of Tree Stand Growth (DMTG)

There are a large number of different models of tree stands and forest communities (see, for example, Oja, 1985; Antonovsky et al., 1992; Shugart, 1986, 1992; Korzukhin and Semevsky, 1992). Nevertheless, problem of construction of appropriate simple dynamic models depending on a few interpretable parameters for description of the main factors only of environment and giving as a result the basic forest taxonomic characteristics remains be actual.

The DMTG (Karev and Skomorovsky, 1998) offers a possible solution. The main variables of the model are: 1) *N*, number of trees; 2) *H*, average height; 3) *D*, average diameter; 4) *V*, average volume of single tree; 5) *W*, growing stock; 6) *O*, wood litter. A model of tree number dynamics is particularly important for exact calculation of the growing stock from the average volume of a tree and the number of stems. Although there are many partly empirical models of tree stand numbers (see reviews by Berezovskaya et al., 1991; Chetverikov, 1989; Korzukhin and Semevsky, 1992), there is no satisfactory theory of tree stand number dynamics.

A simple three-parameter model was constructed by Karev and Skomorovsky (1997). In this model it is assumed that the population can be divided into groups which are mutually disjunct, with different death rates. The qualitative behaviour of the model corresponds to the peculiarities of experimental curves (such as existence of "development phases" and a point of inflection).

Data on trial areas of Scots pine and Norway spruce stands (Advances of Experimental Works . . ., 1964), data on Scots pine plantations of various densities (Rubzov et al., 1976), and also growth yield tables of "normal forest" (Zagreev et al., 1992) of various species, were used for verification of the DMTG model. The mean-square deviation of the model and yield table data was less than 3%. The same accuracy was achieved for more complex variants of the model describing growth of two-species mixed stands.

3.4 The Layer-Mosaic Concept

The layer-mosaic concept (or gap paradigm) is an alternative to the representation of a forest community as a homogeneous formation, traditional in forest science and ecology. The initial idea of the concept, from the beginning of the century, is that the "forest elementary unit" is not an individual tree but an association of trees. The forest community is considered to be a system of spatial mosaics consisting of non-synchronously developing patches, at different stages of development and changing in time as a result of internal dynamic processes. There are three separate basic stages in a cycle of forest renewal: 1) the stage of patch (gap) formation, resulting from a tree or a group of trees falling; 2) the stage of construction, in which young trees dominate; and 3) the stage of maturity, with adult trees. Stability of a community is possible only with a natural combination of the elements of the layer-mosaic structure at different stages of development.

It seems that the layer-mosaic concept arose after works by Aubreville (1971), Dylis (1969) and Watt (1947), and it has developed intensively in recent years. A detailed account of the concept was made, for example, in a monograph by Popadyuk et al. (1994) where its application to deciduous forests consisting of mono- or poly-dominant associations of trees, named "loci" was analysed. A locus is determined by a definite species structure (i.e. density, age, sizes, etc.) at a particular time. The features of development and processes of the formation of loci as patches in "geographical" space were investigated. A verbal model was constructed and then implemented as a detailed computer simulation model (Popadyuk and Chumachenko, 1991).

Another verbal model put forward by Buzykin et al. (1985, 1987), was concerned with research about characteristics of the limiting mode (i.e. whether the regime is stationary, oscillatory or chaotic), that is established in a forest community in the absence of external destabilising factors. This question is one of the basic problems of the study of biological communities and their models. In these papers, the concept of the "cenon" as an elementary structural unit of the forest community was defined (by obvious physical analogy). The cenon is a part of the community of forest plants characterised by uniformity of the area occupied and by uniformity of the generation stage. The patch (or cenon) structure of tree populations was demonstrated and its endogenous dynamics were described, with Siberian spruce as an example.

The basic ideas may be summarised as the following "axioms of the cenon concept":

1. The forest community can be considered as a meta-population of cenons (loci, gaps).
2. Each cenon is a sub-population of trees of one generation, which occupies a fixed area, and arises, develops and perishes as one unit.
3. The birth of a new cenon in the area occurs only after destruction of the old cenon occupying it.
4. The growth of an individual tree depends on interactions with other trees in the cenon, and does not depend on trees belonging to other cenons.

Although the concepts of forest elementary units (gap, locus and cenon) are rather similar, there are qualitative differences between them. "Gap" is a conditional "topographical" cell, i.e. the site of the fixed area in which the birth, growth and death of each separate tree is simulated; "locus" is a real forest structural unit, and its spatial allocation in "topographical" space reflects actual complex processes of renewal and development of forest cover in the released site; "cenon" is a cell not only in "topographical" space but also in "phase" space of forest cover. This means that the trees composing a cenon form a rather homogeneous sub-population, so that the degree of uniformity increases with age and the interactions between these trees are essentially stronger than those between trees belonging to different cenons. The reasons for cenon formation in "topographical" space are the same as for locus formation. The allocation of a cenon as an uniform formation in phase space is caused by the action of "discriminate decay", consisting of sharply increasing mortality of trees with a diameter less then the average in a sub-population. This phenomenon, resulting in formation of a homogeneous sub-population (a cenon), was demonstrated in the work of Buzikin et al. (1985); the appropriate mathematical models of cenon formation were constructed by Karev (1995a, 1995b).

The logical analysis of the verbal model (Buzykin et al., 1985, 1987) has led the authors to the following conclusions:

1. changes in numbers of trees in parts of the area which are far from each other cannot be synchronised for internal reasons. Hence, non-stochastic fluctuations of numbers in some age ranges, which are sometimes observed in large areas, cannot be explained wholly by endogenous reasons;
2. a stable age distribution exists as a limit of endogenous development

of the forest community (distribution of a climax type); and

3. endogenous development and sequence of generations will transform a single age tree stand to an even age one having a stable age distribution as a monotonously decreasing function of age.

From these axioms of the cenon concept, some fundamentally important conclusions for modelling arise:

a. the model of the forest community considers a cenon (locus or gap) as a new individual object, rather than a separate tree;
b. the model of an individual cenon coincides with a model of one generation of a tree population; and
c. the dynamics of a meta-population of non-interacting cenons can be described by an autonomic structural model (see Section 3.6).

The appropriate mathematical methods and models were advanced and investigated in the work of Berezovskaya and Karev (1990), and Karev (1995a, 1995b), in which the theoretical conclusions of Buzikin et al. (1985, 1987) were expressed as exact mathematical statements.

3.5 The Concept of a Single Plant Ecosystem

There is another approach to determine the "minimal forest unit". This is the "single plant ecosystem" concept (Chertov, 1983, 1990) which was briefly discussed in Chapter 1, Section 1.3.6. This concept originated through the influence of process modelling, Uranov's (1965) "phytogenic field" and the theory of primary ecogenetical succession (Razumovsky, 1981). The concept allows the specification of local functioning of a tree-soil system and element cycling in the minimal unit and enables simulation of the effects of spatial heterogeneity of edaphic conditions (both abiotic and biotic) on stand growth and structure. The realization of the concept of individual-based simulation has shown its suitabilty for forest modelling (Chertov, 1990; Chertov and Komarov, 1995, 1997; Chertov et al., 1997). A mosaic structure of a forest ecosystem is always obtained as the output from running such a model.

3.6 Gap-Models of Forest Communities

Gap-models have been widely used in the 25 years since the first JABOWA model (Botkin et al., 1972). Monographs by Botkin (1993) and Shugart (1984, 1992) contain the basic concepts, models and results of gap-modelling.

From the standpoint of theoretical ecology, gap-models represent realisations (more or less simplified) of the layer-mosaic concept. Models of the latest generation (FORET, FORSKA, ZELIG) are distinguished by a considerably more extensive use of theoretical ecology to justify them, as compared with the first gap-models. On the other hand, gap-modelling may be considered as a large-scale computer experiment to verify the main principles and consequences of the gap-paradigm.

A model of separate gaps describing the dynamics of trees on sites of specified area is the basis of every gap-model. Each tree of a given species is characterised at any one time by a certain set of variables: height, diameter, etc. The equation for growth depends on light, temperature and other environmental parameters; competition for resources can also be included. In the vast majority of gap models, the influence of other gaps is not taken into account. Tree establishment and mortality on a site is described by stochastic processes; the initial distribution of trees is also stochastic. As the model is stochastic, forecasts are calculated as the average of a rather large number (80–100) of independent solutions.

In the work developing the layer-mosaic concept (see Popadyuk et al., 1994), much attention is given to parameters of patches, mainly to their spatial size. Various tree species require gaps having species-specific minimal sizes for normal regeneration. External factors, such as fires, insect attacks, felling and so on, can have an important influence on the formation of patches, and on their size and species composition. The consecutive implementation of the rules of the layer-mosaic concept requires the creation of simulation models that are much more detailed than the usual gap-model. One such model (Popadjuk and Chumachenko, 1991) considered the three-dimensional structure of a forest community and operated with three-dimensional spatial cells of dimensions 5 × 5 × 10 m.

The size of a patch in a gap-model is usually set equal to 10 × 10 m. Apparently, this is approximately equal to the area of the patch formed where a large tree has fallen down, corresponding to a "main part" in the real mosaic of multi-scale patches of a forest community. The dependence of species composition of a patch on its size, mentioned above, is partly simulated by elimination of the pioneer species from the list of possible species determined by given environmental conditions. Computer experiments on the size of separate gaps in the ZELIG model were carried out by Urban et al. (1991).

The size of patches is probably not decisively important for a population of a single tree species and parameters of patches are, therefore, not included in the cenon models. The choice of gap sizes in a gap-model is the important problem and must be a compromise. Modelling of large patch dynamics must take into account the natural consequences of gap size differences and consequent asymmetric competition. On the other hand, interaction and competition processes between different patches must be included in the model if patch sizes are small.

Some gap-models (e.g. Urban et al., 1991; Popadyuk et al., 1994) take into account processes of interaction with trees in the neighbouring patches (e.g. competition for light). The majority of models assume that there are no significant interactions between trees belonging to different gaps. Calculations with the appropriate models confirm this assumption, namely: 1) attainment of satisfactory forecasts of the dynamics of various forest communities, from boreal to tropical, when no account is taken of interactions between gaps; 2) the absence of interactions leading to rapid movement of the gap ensemble distribution towards a stable situation; this convergence can be observed in computer experiments with the model and is supported by analytical models (see below). On the other hand, a sufficiently strong interaction between patches may force synchronisation of their development. Thus the system as a whole may have an oscillatory mode not observed in natural forests, which are not subject to broad-scale influences.

3.7 The Problem of Expansion of Spatial and Temporal Scales and the Creation of a Mathematical Theory of Gap-Modelling

Research on the influence and long-term consequences of global climate change and anthropogenic impacts on forest vegetation requires a transition to long time scales and should include study of asymptotic behaviour of the model with various environmental scenarios. The problem of the necessary enlargement of the spatial and temporal scales of local gap-models is a very important and difficult one. Reviews of relevant results have been published (Shugart, 1984; Solomon and Shugart, 1984; Urban and Smith, 1989; Urban et al., 1991).

The enlargement of spatial and temporal scales conflicts with the individual orientation of gap-models that trace the dynamics of each separate gap (and even of each separate tree in a gap), so that the spatial scale of the model is limited by the capability of the computer. Enlarging the time scale to take the dynamics of a large number of generations into account gives rise to the following problem. The output of the landscape gap-model is a map of a forest ecosystem represented by a mosaic of gaps, with certain coordinates and a range of states. The stochastic character of the processes of tree birth and death, with a random initial distribution, leads to the result that the observed and calculated states for each separate gap will be quite different (and so will the map as a whole), as a consequence of the stochasticity of the process. Thus, the major advantage of gap-models – "individual orientation" – is also a major constraint to enlargement of the spatial and temporal scales. On the other hand, the statistical characteristics of the map as a whole, e.g. age distribution, average height and diameter of trees, standing crop of wood etc., can be close to the real values.

It seems to us that these problems cannot be solved completely by using better programmes and more powerful computers, but only through new mathematical approaches and methods. The basic direction is prompted by statistical dynamics, which do not trace a trajectory of each separate object but move on to the study of assemblages. Some such methods have been based on the theory of structural models of forest dynamics (Berezovskaya and Karev, 1990; Karev, 1993b, 1994), an approach which may be considered as a mathematical theory of gap-modelling, described below.

3.8 Individual-Based or Structural Models of Populations

The basic mechanisms which govern behaviour and dynamics of a population such as birth and death rates are directly connected to individual differentiation. The greatest interest in modelling the dynamics of tree stands for management lies in the values of growing stock, age structure, average height and diameter, etc. (and their variances). Thus, both for building realistic models and for getting the required information about the state of a simulated community, it is necessary to pass from an amorphous model of population number to models which are structured by an internal variable describing the state of separate individuals. The theoretical aim of such models is to describe population behaviour on the basis of individual behaviour.

The dynamics of tree populations for periods longer than the lifetime of one generation describe the renewal processes, and usually take the age structure into account. Analysis of the age dynamics (without considering any characteristic of the individuals other than age), starting from the work of Lotka (1925) and Volterra (1931), has been done on the basis of linear models of three types: both time-discrete and age-discrete models; time-continuous and age-discrete models (Leslie's models); and, finally, both time-continuous and age-continuous models (Lotka's models). Models of these three types have been applied to study the asymptotic behaviour of populations and have also been applied to the modelling of forest dynamics (see Korzukhin and Semevsky, 1992, for some examples). Thus, the main result from Lotka's models is given by the Lotka-Sharp theorem about the existence of a unique limit for stable distribution of the population; the modern formulation of this result was given by Webb (1984).

Non-linear models of age-dynamics are more complicated to analyse, but are much more realistic. Thus, non-linear, age-discrete models (Gavrikov et al., 1985) were constructed to study a well-known phenomenon of forest dynamics, namely variation of the age structure of the population of trees regarded as strong dominants. Such models have served as a basis for modelling influence of insects on the age structure of tree populations (Antonovsky et al., 1991a, 1991b; Kuznetsov et al., 1996).

Great interest has been shown in models of age dynamics with continuous age and time after the work of Gurtin and MacCami (1977), which was followed by a flood of publications on analysis and applications of non-linear models of age structure with respect to many internal structural variables that describe the dynamics of a separate individual, such as biomass, size etc.

Structural models of a population have two scales of description: individual (i) and population (p). Let the individuals of a population be characterised at any one time by age a and finite number n of variables $(x_1, .. x_n) = X$, where X is called the i-state. At the individual scale, a model must specify a) the rate of change of the i-state, (that is by a dynamic system of equations); b) the death rate; c) the birth rate and specification of the distribution of the new individuals; and d) the dependence of these rates on the i-state (if it exists), and the prevailing environmental conditions and state of the population as a whole.

As a model at the individual level exists it is possible to derive balance laws generating the time evolution at the population level. Population state is given by the density $l(t,a,X)$ of number of individuals which have at t moment an *i*-state X and age a. At the population level a model amounts to a specification of a) initial density $l_0(a,X)$; and b) the number of new individuals per unit time $B(t,X)$, calculated from the birth rate and age distribution.

The dependence of growth, birth, and death rates on the external conditions, general state of the population, and interactions among individuals, may be described formally in structural models by so-called regulating functionals (Tucker and Zimmerman, 1988), named also *generalised variables*, that are specific population averages. Examples of generalised variables are the number of individuals or biomass of a population. A quantity of external resources per individual can depend from these values. If the model does not depend on any generalised variable, then it is called an autonomous model.

Thus, a structural model of a population consists of three linked blocks: a dynamic system for *i*-state change, i.e. for the dynamics of each individual; the Kolmogorov forward equation, well-known from the theory of random processes, describing the *p*-state change, i.e. the dynamics of population density, and the integrated boundary condition describing the process of regeneration. A good introduction to the theory, with numerous examples, is given by Deickman and Metz (1986).

The main goal of a structured population model is to study evolution with time of the initial density, $l_0(a,X)$ and, in particular, the asymptotic behaviour of population density, $l(t,a,X)$, when t is very large. For autonomous models, it has been shown (Karev, 1989, 1993a), that there is a constant growth rate of populations, λ, such that the normalised density of a population, $l(t,a,X)exp(-\lambda t)$, converges (when $t \rightarrow \infty$) towards the density of the unique stable distribution. Thus the speed of convergence from any initial distribution is exponentially rapid. The exact form of the density of the limiting distribution was found, and a convenient method for its calculation was offered by Karev (1993a, 1996).

As mentioned in Chapter 2, the theory of plant competition is a separate and important part of plant ecology. Mathematical model of interaction and competition between separate trees for external resources (light, water, area of growth, etc.) is a necessary link to proceed from the models of individual trees to the models of tree population.

A widespread approach, appropriate for analytical models, is to describe interaction processes by calculating the external resources available to an individual tree. The results obtained are then described with the aid of generalised variables. Typical examples of this approach are models describing competition for light (the leaf canopy usually being simulated using the random turbid medium concept (e.g. Monsi and Saeki, 1953; Kull and Kull, 1989), with some modifications for poly-dominant and multi-layer stands (e.g. Karev, 1984, 1985a, 1985b, 1985c; Korzukhin and Ter-Mikhaelian, 1987, 1995); models analysing competition for area of growth (e.g. Galitsky and Tuzinkevich, 1987; Galitsky and Komarov, 1987; Korzukhin and Semevsky, 1992); distributed models of root competition (e.g. Karev and Treskov, 1982; Tuzinkevich, 1988); and some others. Some alternative approaches and references are given in reviews by Antonovsky et al. (1991) and Karmanova et al. (1992).

Synthesis of the models describing dynamics of a separate tree and the models analysing competition for external resources makes it possible to proceed to population models.

3.9 Structural Models of Meta-populations

The further development of the theory of plant growth and competition has been construction and research of structural models of meta-populations (Berezovskaya and Karev, 1990; Gulpin and Hanski, 1991), i.e. models in which the "individual" is a sub-population consisting of simpler individual units. This new theory has recently developed in two different directions: modelling of mobile individual meta-populations (Cyllenberg and Hanski, 1992, Cyllenberg et al., 1997), and modelling of fixed individual (plant) meta-populations and communities (Berezovskaya and Karev, 1990, Karev, 1992, 1994).

Structural models of meta-populations have three scales of description: individual (*i*), sub-population (*sp*) and meta-population (*mp*). The individual scale is the same as in structured population models. At the sub-population scale, a model amounts to specification of a) interactions among individuals in a sub-population, formulated in a dynamic system of equations; b) the death rate of a sub-population as a whole; c) the birth rate of new sub-populations and specification of their initial numbers and of the distribution of the set of *i*-states (where *i*-state is a description of the sub-population); d) the rate of change of sub-population numbers and the way their distribution changes; and e) the dependence of all rates on the current state of the sub-population and on the prevailing environmental conditions, and the state of the metapopulation. The description of a meta-population is the same as for a population, but with the "individual" replaced by the "sub-population".

Structural models of tree populations, in which the individual is a separate tree, have been put forward repeatedly, but the complex competitive interactions between trees has resulted in essential non-linearity of such models. Thus these models do not permit qualitative solutions and are even difficult for computer implementation.

According to the layer-mosaic concept, or gap-paradigm, the situation essentially varies with transition to the new individual object, the sub-population of trees (i.e. locus, cenon or gap), and consideration of a tree population as a meta-population of cenons or gaps. We will consider briefly the structural models of one generation of populations, as such models have independent interest.

Taking account of interactions amongst trees, described by appropriate generalised variables, complicates the investigation and application of models. These difficulties may be overcome in some classes of models of theoretical and practical interest, which are defined by additional conditions of the initial or current distribution of sub-populations of: 1) a single-species, even-aged tree stand (a sub-population with uniform distribution of size and species); 2) a multi-layered tree stand (i.e. non-uniform distribution, e.g. a population divided into m groups with identical trees within each group); 3) a mixed stand practically the same as a model of a many-layered tree stand (but the equations describing dynamics of the *i*-th species can depend on its number); 4) a tree stand with a narrow *current* distribution of sub populations (with moments greater than 2 equal to zero); and 5) a tree stand with a narrow *initial* distribution. Models 1) to 3) can be used when the initial distribution can be approximated by the piecewise constant function, whereas models 4) and 5) are especially useful for modelling gap dynamics construction. The essential feature of models 4) and 5) lies in dividing the initial model into two parts. The first part allows generalised variables to be taken into account and gives a result that is used to describe the dynamics of a separate gap. The first stage model really describes characteristics of the main population dynamics and is interesting in itself. The methods of construction of these non-uniform population models, with applications to forest communities and specific examples, were given by Karev (1995a, 1995b).

3.10 Enlarging the Spatial and Time Scales and Meta-population Forest Models

The first stage of spatial scale enlargement of gap-models is the transition from a model of a homogeneous tree sub-population (i.e. in a cenon) to a model of a heterogeneous sub-population. The basic difficulty is to take into account of the processes of interactions amongst trees, i.e. dependence of the growth equations on regulating functionals. Sub-population models of one generation may overcome this mathematical difficulty with an appropriate choice of initial distribution, or some of the assumptions about distribution of tree properties considered in Section 3.9.

In the models constructed by these methods, the interactions amongst trees turned out to be confined within a gap, so that the gap models can be rather complex. However, the structural model of a gap meta-population proved to be an autonomous one, on the basis of the assumption that it is possible to neglect any interactions between gaps. This circumstance is important, as it enables the asymptotic behaviour of all structural autonomous models and some of their generalisations to be investigated completely.

The second stage of spatial scale enlargement of forest community models is concerned with the modelling of meta-populations consisting of large numbers of gaps. Dynamics of a gap meta-population can also be considered on different temporal scales (i.e. over one or many generations). We emphasise that the model of one generation of a gap meta-population has the same formal structure as a model of one generation of a tree population. Thus it is necessary to describe models of gap dynamics and the processes of gap interactions (similarly to interactions amongst trees) and to define the rates of their birth and death as entire objects. The results occur to be essentially depended on the kind of initial or current tree distribution (see Section 3.9) in a sub-population model.

Models of meta-populations can be constructed by the same methods as sub-population models of classes 1) to 5) in Section 3.9. Of these, the models of heterogeneous meta-populations similar to sub-population models of classes 4) and 5), are the most interesting. A new and important conceptual peculiarity of these models is the appearance of two different distributions – "internal" one of the trees inside a sub-population, and "external" one of the sub-populations as individual objects of a meta-population. The "internal" distributions are not supposed to be identical for all sub-populations.

These models allow the study of dynamics of spatially non-uniform tree populations, remaining within the framework of the ordinary differential equation based models (although much more complex models based on partial differential equations are usually applied to such situations). The models of this class represent an adequate mathematical tool enabling the construction of models of meta-populations (i.e. forest communities) on large spatial scales on the basis of "local" models (i.e. gap-models).

In the third stage of spatial scale enlargement, it is possible to research the asymptotic behaviour of a model over long periods of time. As all interactions between trees take place within a gap, the gaps being considered as non-interacting formations, the appropriate structural model of meta-populations turns out to be autonomous.

The general result is that, irrespective of the details of the description of the dynamics of an individual tree, and of the processes of competition and interaction of trees within sub-populations, there is a unique stable distribution for the meta-population as a whole, defined by the limit of the convergence process according to the exponential law, while originating from any initial state (Karev, 1993a). The precise form of this limit distribution has been found to depend on the initial structural values, birth and death rates of sub-populations, as well as on the parameters describing the environmental conditions.

In many cases, exact expression for the limit distribution is not required, since only the basic characteristics, such as growing stock or average height, are important. The values of all these *generalised variables* in a limiting stable state have been defined in precise form.

Earlier results have led to new methods of applying gap-models over long intervals of time and making analytical studies, using these models, of the influence of different environmental variables. By extracting the analytical kernel from a computer gap-model (i.e. a model of local gap dynamics, with rules for birth and death of gaps and their initial distribution), it is possible to define the limit distribution of a meta-population and to find limit values of the necessary generalised variables *without* making a computer accounts of a transitive modes.

Moreover, it is possible to investigate *directly* the dependence of all generalised variables and the asymptotic distributions on various parameters, included in the model (e.g. the initial distributions, etc.). This investigation corresponds to the analysis of climax states of forest community under various scenarios of climatic and anthropogenic influences. The research can be done, using available software, by qualitative analysis of the systems of differential equation.

3.11 Dynamics of Forest Areas

3.11.1 Statement of a problem

The above approach describes asymptotic behaviour of the forest meta-population, making two assumptions: 1) that the forest vegetation already has a mosaic structure in its initial state; and 2) all gaps belong to a single type of forest cover, which is defined by species structure and site conditions.

Both assumptions appear too restrictive. Thus, if the initial structure of vegetation was fairly homogeneous (for example, after rapid colonisation of an area, as a result of catastrophic external influences, such as fire, insect attack, felling, etc.), the formation of small-scale structure occurs only after several generations. Therefore the dynamics of the forest vegetation can be simulated on larger spatial scales for areas which are subject to strong external influences for about the duration of one generation. This aim can also be considered within the framework of structural models of the dynamics of forest areas (Karev, 1994).

The following circumstances are part of this problem. Firstly, a model of the forest community from the point of view of the gap-paradigm coincides with the model of a meta-population with several different gap types. Secondly, on this assumption, all gaps in the meta-population model are of the same size. Thus the number of gaps having the given age and value X of the structural variable at a given time is proportional to the area occupied by forest vegetation of the same age and X value. In this case, descriptions of forest community dynamics, in terms of areas and distribution of a metapopulation with several types of gaps, differ only in interpretation.

3.11.2 Structural model of succession

Mathematical modelling of plant succession has been developing in two essentially different directions (Shugart, 1984; Liu and Ashton, 1993). The first is the gap-modelling of succession, i.e. models of the JABOVA-FORET type (Botkin,1993). The second involves Markov models, which operate with proportions of areas occupied by certain stages of development of the succession, but without taking into account the inter-stage dynamics (Horn, 1975; Shugart and West, 1981; Shugart et al, 1981; Cherkashin, 1981). The mathematical tools required for these two types of modelling approaches are quite different.

Modelling of succession processes requires the construction of a landscape gap-model to run over long periods of time (hundreds of years) under various environmental scenarios. But this enlargement of spatial and temporal scales conflicts with construction of gap-models describing the dynamics of each separate gap. So a model of succession covering a range of spatial and temporal scales is limited by computer resources.

Synthesis of these two main types of succession models is a current problem, that has been approached by a number of researchers simultaneously from different points of view. A computer simulation model using linkage between these two approaches was constructed by Acevedo et al. (1995). A mathematical feature of this model is the use of a semi-Markov process for description of the replacement process of one gap by another, i.e. caused by the significant spatial size (about 1 ha) of the elementary succession object in this model and, therefore, the necessity for description of processes that are transient in time.

An alternative approach is based on the idea that the elementary object of vegetation is the area occupied by one stage of a single succession (Razumovsky, 1981). Because this area can be very small (less than 1 m^2) the duration of transient change of such units may also be very small. This assumption allows construction of a succession model within the framework of differential structural models.

An analytical approach to this problem of synthesis was taken by Karev (1994, 1996), who constructed a structural model of succession and biological community dynamics. This model allows the description (at a phenomenological level) of the natural processes of succession dynamics of forest communities within a given area, together with, and in relation to, internal dynamics. Specific cases are represented by Markov models and gap-models in the form of analytical structural models (and also by models of populations with both age and functional structures and a complex life cycle).

The values to be defined in a structural succession model are the areas $S_i(t,a,X)$, occupied at each moment t, by a forest of certain types i, age a, and state X. The concept of forest type includes species composition, site class, etc. Succession age a means the time since the appearance of the forest in a given area (this is generally not the same as the age of the tree stand, which is the average age of the dominant species). The forest state at moment t, can be defined by current values of basic stand parameters (e.g. height and diameter of trees of a given age and species, and the stand density). The model of state dynamics is considered to be given and it is assumed that both the rate of area decrease (caused by self-thinning, fire, pests, felling, etc.) and the stochastic succession matrix of transition, specifying the succession model, are given. Elements of the succession matrix of transition are equal to the probabilities that a unit of area vacated by a type j forest will be replaced by a type i forest.

For the structural models of this kind, it has been shown that the natural dynamics of the community lead to an unique stable state (i.e. distribution of the areas $S_i(a,X)$, occupied by forests of defined types, age and status), which is attained at an exponential rate (Karev, 1994). In the stable state there are different processes of growth, mortality, change and renewal of vegetation cover on each fixed site, but the distribution of areas of the forest ecosystem remains static.

3.12 Succession and Ergodicity

The "ergodic properties" of a system are well-known in physics as a result of its many possible applications. The application of ideas based on ergodic theory to biological problems has long been the subject of discussion. The "ergodic hypothesis in ecology" was formulated exactly by Molchanov (1975) as follows: the areas S_i of the biocenoses making up a successional sequence should be proportional in a climax state of association to the times, T_i, of their development in the succession. Thus

$S_i / T_i = K = constant$ for all $i = 1, .. n$ (12.1)

More precisely, the time T_i represents the average time of renewal of the *i*-th biocenose in the climax state.

The importance of the ergodic hypothesis is that the *area* distribution can be used for estimating the current state of development in *time*. This is especially useful in forest ecology for study of the succession processes, the duration of which is normally much longer than the human life span.

Equation (12.1) allows quite simple estimation of the deviation of the observed state of a forest ecosystem from the theoretically stable state. The area and "proper" times of the biocenoses are put on the axes of coordinates. Then the ergodic hypothesis asserts that, in a climax state of association, the points (S_i, T_i) will lie on a straight line through the origin. Hence, it is possible to estimate the difference between the observed and climax system states, from the deviation of these points from the line.

It seems that experimental foundation of the ergodic hypothesis may occur to be very difficult because of duration of forest succession processes. So a theoretical research of the ergodic properties of ecosystem climax state and mathematical methods of calculation the ergodic constant K and the proper times seems to be especially important.

This research has done within the frameworks of a mathematical structural model of succession, described above. The general ergodic theorem (Karev, 1997) asserts that the equalities (12.1) are valid not only for areas, but also for any generalised variable $Q_i(t)$, that is

$Q_i / TQ_i = K$ (12.2)

Here Q_i are the values of generalised variable in the climax state of association that appear as specified averages determining limit area distributions $S_i(a,X)$, and TQ_i is the corresponding "proper time" (which may be calculated within a structural succession model). The constant $K = V/m$ is universal for the given model and is equal to the common area of succession system V, divided to the mean time of renewal of the succession system at the stationary state m.

Note that practically all the relevant characteristics of the forest ecosystem, such as average tree height and diameter, distribution of areas by groups of age, growing stock etc., can be expressed as the appropriate generalised variable.

Equation (12.2) is the foundation of theoretical simple *method of complex estimation* of the deviation of the observed state of a forest ecosystem from the theoretically stable state. Let postpone the values of any generalised variable Q_i for every i-th components of association in its climactic state on the y-coordinates and the values of corresponding proper times TQ_i on the x-coordinates. Then the points $(Q_i\ TQ_i)$ for all i will lie on the *common* straight line for *every* generalised variable.

The value

$$D^2[Q] = \sum_i (Q_i(t)/[TQ_i] - K)^2$$

where $Q_i(t)$ be the observable values of the *given* generalised variable for i-th biocenose at t moment, may be taken as a measure of the deviation (by *characteristic* Q) of an observable state of the forest community from the equilibrium climax state.

Hence, it is possible to find out which characteristics of the real forest community (e.g. stock of wood, or age structure) deviate most from a theoretical steady state, and to estimate the degree of deviation. This is especially valuable for estimating the status of reserve land.

The importance of the generalised ergodic theorem lies also in the inverse statement: from the validity of equation (12.2) for all generalised variable, it follows that the succession system is in the climax state.

The role of rapid stages and small biocenoses also becomes clear. Their importance is not immediately apparent in a climax association: if the regeneration time of a biocenose is small, then the occupied area (and the total biomass, etc.) is also small. However, such a biocenose can be a transition stage between biocenoses with long regeneration times. Thus the removal or destruction of a rapid stage or small biocenoses can result in the destruction of an association of biocenoses as a whole; and the maintenance of the existing biodiversity can be a necessary condition for the existence of the association.

3.13 Some Perspectives

Analytical models describe the dynamics of a few essential variables and parameters, and their qualitative features as well as facilitating study of limit regimes and changes in spatial boundaries. It seems promising to use both analytical and simulation approaches, such that analytical models are regarded as sub-blocks of the simulation models. This makes it possible to adapt models to the modelling objects, to forecast achievable situations, and to assess the boundaries of valid applications and the precision of the simulation approach. On the other hand, analytical models can be a separate, useful (and often unique) tool for investigation of the qualitative behaviour of complex ecological systems over a long time span.

One of the most important aspects of the application of mathematical modelling is the study of the behaviour of systems near the "dangerous boundaries" of both parameters and variables and, in this connection (probably, catastrophic) reorganisation of the functioning system (Bazykin, 1985). It is known that the "critical modes", arising when the parameter values of a dynamic model approach the "dangerous boundaries", are described by only a few of the main parameters and variables. Thus analysis of a functioning of a complex multi-dimensional system close to its critical limits reduces to study of the qualitative behaviour and bifurcations of an essentially simpler "low-dimensional" system.

Research experience of the qualitative behaviour of "low-dimensional" models of complex systems has resulted in an appreciation of the opportunity to define and implement the inverse role, namely, to construct a portrait, or kinetic model, that is the simplest dynamic model in which all the observable kinetic types (or phase portraits) of the system are realised (Berezovskaya and Karev, 1997). Thus the quantitative aspects of modelling and the problem of validation with experimental data become less important.

The aims of explanatory and portrait-kinetic models are complementary. Explanatory models are intended to answer the question: what are the kinetic properties of a system with a given structure? Portrait-kinetic models answer the inverse question: what is the structure of the model with given kinetics? An interesting and important example of a portrait-kinetic model, using the analysis of natural and experimental data, is a model of the dynamics of forest insects (i.e. a phytophage-entomophage system) by Bazykin et al. (1995). We can now examine the possibility of synthesis of the basic directions of mathematical modelling, explanatory and portrait-kinetic, from the following point of view.

The next general problem is the description of complex system dynamics. Long periods of predictable, steady, determined (by Laplace) development can suddenly be replaced by rather short periods of unpredictable, seemingly random and probably catastrophic changes of the system dynamics. During such periods, small changes of external environmental variables and of internal biological parameters can lead to qualitative, bifurcating reorganisations of system status and dynamic modes.

Apparently, it was only in this century that it was fully realised that the occurrence of seemingly unpredictable reorganisation modes cannot be explained as a whole by random perturbation of Laplacian dynamics. On the contrary, there are qualitative dynamics of another kind, such as the theory of catastrophes, and its mathematical tool, the theory of bifurcations of dynamic systems, which were developed for investigation of this particular kind of dynamics. Thus the complete description of complex systems dynamics (at our current level of understanding) requires synthetic approaches of the following kinds.

a) Complex dynamic models (both simulation and analytical) with many variables and parameters, and rather simple qualitative behaviour, can be used for the description of deterministic phenomena. Good examples of a suitable class of analytical models are the structural models of populations and communities; an interesting example of the result is an ergodic hypothesis in biology.

b) For the description of catastrophic reorganisation events, it is necessary to allocate a few essential variables and parameters, to construct the simplest model having observable or expected stereotypes of dynamics, to construct the "hull" of these stereotypes, i.e. the whole complex of dynamic modes realised in the model, and to investigate the conditions of change of modes composing this complex.

It is difficult to place these model types, with various characteristic time scales, into the hierarchical system of models. However, the problem has become topical, because of the weakness of scientific forecasts of social-economic developments and consequent mistrust of these forecasts. The methodology now offers choices of alternative paths of development in the approaching conditions of ecological and population explosion.

4 Conclusions

Modelling of the forest stand, of forest growth, and of forest ecosystems, is currently developing intensively and there have been considerable achievements in forest ecology and in practical applications. From a theoretical point of view, modelling approaches have led to the development of a theory of forest succession (West et al., 1981; Shugart, 1984; Shugart et al., 1992a; Karev, 1993). There are also many applications of forest models in forest ecology and forest science. Models were first used for quantitative prediction of traditional forest practice, e.g. the effects of felling on stand dynamics and composition. Nowadays there is a discussion on their use as a new tool in forest management (Coates and Burton, 1997; Battaglia and Sands, 1998). However the FORCYTE model only (Kimmins and Scoullar, 1979) is an excellent pattern of user-friendly interface.

Recently, the models have been used for evaluation of the impacts of pollution (Chertov, 1990; Krieger et al., 1990; Arp and Oja, 1992), of insect attacks (Shugart and West, 1977; Isaev et al., 1984; Bazykin et al., 1995) and forest reclamation of industrial badlands (Chertov et al., 1998b). Currently they are widely used for prediction of the long-term effects of expected climate changes on forest ecosystems and the forest biome as a whole (Post and Pastor, 1990; Shugart et al., 1992b; Post et al., 1996; Smith, 1996; Lindner et al., 1997; Bugmann and Cramer, 1998). The process of model evaluation against experimental data is also started (Smith et al., 1997; Aherne et al., 1998; Walshe et al., 1998).

The material presented in the review allows clear conclusions to be drawn about some of the general prospects for forest modelling development in the near future. The points of growth are considered to be the following:

(i) in simulation modelling: a) combination models; b) spatial interactions; c) tree ontogeny; d) spatial edaphic heterogeneity attributable to relief, water, and soil; and e) regional modelling.

(ii) in theoretical analytical modelling: a) theory of gap formation; b) synthesis of Markov type and gap-models of forest succession; c) the study of "macro-characteristics" of forest landscape models, depending on "micro-characteristics" of separate gaps and trees, using

an individual-based approach. Another very important problem is: d) modelling forest ecosystem dynamics near the "dangerous boundaries" of external factors.

(iii) a separate and important prospect is the creation of a system of forest models with different approaches, spatial and temporal scales.

There is understanding in the West of the necessity to create combination models unifying existing approaches in individual-based and process modelling (Huston et al., 1988; Bossel et al., 1991; Levine et al., 1993) but no clear understanding of the idea of real ecogenetical development of the tree-soil system in the course of succession (primary or secondary). The development of stand and other biotic components of the ecosystem, both natural and artificial, is always accompanied by soil development (primarily the accumulation of organic matter and nutrients). This is a forgotten main postulate of Clements' theory, recently developed by Razumovsky (1981) and by Chertov and Razumovsky (1980). There have been no really successful attempts at simulation of such dynamics, even in cases in which tree growth and soil changes have been considered.

However, there are serious problems in the transition from gap models to combination models. One of the more obvious difficulties seems to be the elaboration of the ecological parameters of the tree species (silvics) so that they are suitable both for modelling the dendrometric characteristics of trees and stands, and for modelling the biological cycle in the forest ecosystem. An additional, and maybe unexpected, problem is a change of silvics with tree age, i.e. there is a shift in values of tree parameters at different ontogenetic stages (Smirnova et al., 1990). Special efforts are required to quantify the parameters for trees with free growth but this is a special case

There are also spatially related modelling problems. For individual-based models, the main problem is the description of spatially dependent interactions between trees. Two approaches to this problem may be distinguished: firstly, description of local interactions for each tree within a stand; secondly, a global description of competing processes reflected by the spatial structure of the community. This largely reflects the redistribution of resources amongst the trees. These approaches may be applied to modelling at different spatial and temporal scales. Local description is necessary for analysis of the basic mechanisms determining forest site dynamics and the rules for coexistence of various spatially distributed sites (gaps, patches, cenons). Global description may be used for diagnosis of the effects of competition on spatial structure and development of forest stands, particularly utilising remotely sensed data. It should be noted that there has been a significant shift of interest in studying spatially explicit structures in different scientific areas and in applied mathematics and this well help to improve forest models. A change of terminology from "individual-based gap" to "patch-based spatial" modelling (Smith, 1996; Wu and Levin, 1997) reflects an increasing role of spatially explicit approaches nowadays.

Simulation modelling at landscape and regional scale is currently in the process of development. The progress of these types of forest models is of great importance and we expect an increase in the amount of research in this field. There are currently two methodological approaches to regional modelling: *(i)* the application of gap-modelling methodology to large areas, using the increased capability of modern computers; and *(ii)* the use of various Markov chain type models for estimation of only the most general characteristics of landscape dynamics. In addition, there is now a new approach, the creation of synthetic structured models of succession and landscape dynamics.

A hierarchical system of forest models with characteristic spatial and temporal scales is likely to be an important next step for further development in this field. There are some examples of such a system already in existence (Karev, 1993, 1995a,b). More significant and valuable systems may be constructed on the basis of the DMTG model and the SPE model, and their appropriate modifications, considered earlier. The construction of model systems offers the principal opportunity for joint application of simulation and analytical approaches. The realms of application of these approaches have different characteristic time scales. Analytical models are most effective, with research into asymptotic behaviour of a system over a long period, in stable environmental conditions, and with no site/landscape variability. Such models are intended to answer the question as to what stable state or dynamic mode will be established in the system if existing external conditions and tendencies are maintained, and also to describe the internal structure and functioning of the system.

Simulation models are the best tool (and sometimes the only tool) for study of transitional system dynamics between initial and asymptotic states in a variable or changing environment (i.e. soil, landscape, climate, anthropogenic impacts).

Logically the advance of different model approaches should lead to simultaneous development of comprehensive model systems that unify theoretical, ecological, silvicultural, industrial, economic, decision-making and social modelling. This is a target for the near future. At the same time, another approach will involve consistent building up of a database and flexible application of different forest models for solving various specific silvicultural, environmental and economic problems.

At present, simulation modelling is largely an art (Parton, 1996), based on the knowledge and intuition of the modellers. Thus forest ecosystem modelling should be a practical tool for converting the art to a precise science in the near future. There is a necessity for much intensive research, both in experimental science, as initiated by the responses predicted by modellers, and in the creation of new approaches in the modelling itself.

In considering the development of forest ecosystem modelling, we can predict that it will be a new tool in sustainable forestry, and will replace the "yield tables" which have been the basis of silviculture since the 19th century. These tables are based on the concept of environmental stability, primarily on constancy of site conditions and of forest growth. However, we live in a changing environment, and the idea of forest growth as a process with a site-dependent fixed rate is usually wrong (Chertov, 1981; Sennov, 1984). Moreover, the expected changes of climate are likely to increase the ineffectiveness of existing yield tables for stand growth prediction. Thus the use of forest ecological models, which are sensitive both to changing environmental factors and to competitive interactions of the trees, will become more valuable for prediction of stand growth and development, and will include consideration of the development of other biotic and abiotic ecosystem components.

References

Aber, J.D. and Melillo, J.M. 1982. FORTNITE: a computer model of organic matter and nitrogen dynamics in forest ecosystems. *University of Wisconsin Research Bulletin R3130.*

Acevedo, M.F., Urban, D.L. and Ablan, M. 1995. *Landscape scale forest dynamics: GIS, gap and transition models.* In: Goodchild, M.F., Stewart, L.T., Parks, B.O., Crane M.P., Johnston C.A., Maidment, D.R. and Glendidning, S. (eds.). GIS and Environmental Modelling. Progress and Research Issues. GIS World. Fort Collins, Colorado. Pp. 181–185.

Addiscott, T.M. 1993. Simulation modelling and soil behaviour. *Geoderma* 60: 15–40.

Advances of Experimental Works in Timiriazev Experimental Forest. 1964. Proceedings of Timiriazev Agricultural Academy. Moscow. 519 p. (In Russian).

Ågren, G.I. and Bosatta, E. 1987. Theoretical analysis of the long-term dynamics of carbon and nitrogen in soils. *Ecology* 68: 1181–1189.

Ågren, G.I. and Bosatta, E. 1996. *Theoretical Ecosystem Ecology – Understanding Element Cycles.* Cambridge University Press. Cambridge.

Ågren, G.I., McMurtrie, R.E., Parton, W.J., Pastor, J., and Shugart, H.H. 1991. State of the art of models of production decomposition linkages in conifer and grassland ecosystems. *Ecological Applications* 1: 118–138.

Aherne, J., Sverdrup, H., Farell, E.P., and Cummins, T. 1998. Application of the SAFE model to a Norway spruce stand at Ballyhooly, Ireland. *Forest Ecology and Management* 101: 331–338.

Aleksandrov, G.A. 1992. *Formalization and analysis of the succession system properties.* In: Logofet, D.O. (ed.) Mathematical Modelling of Plant Populations and Phytocoenoses. Nauka. Moscow. Pp. 30–37 (In Russian).

Alekseev, A.S. 1993. Size structure of tree plant populations: its principal types, mechanisms of formation and use in theoretical population analysis. *Journal of General Biology, Moscow* 54: 449–461. (In Russian with English summary).

Antonovsky, M.Ya., Berezovskaya, F.S., Karev, G.P., Shvidenko, A.Z. and Shugart, H.H. 1991. *Ecophysiological Models of Forest Stand Dynamics.* WP–91–36. IIASA. Laxenberg, Austria. 97 p.

Arp, P.A. and McGrath, T.P. 1987. A parameter-based method for modelling biomass accumulation in forest stands: theory. *Ecological Modelling* 36: 29–48.

Arp, P.A., McGrath, T.P. and Beck, J.A. 1987. A parameter-based method for modelling biomass acumulation in forest stands: an application. *Ecological Modelling* 36: 49–64.

Arp, P.A. and Oja, T. 1992. *Acid sulfate/nitrate loading of forest soils: forest biomass and nutrient cycling modelling.* In Grennfeld, P. and Thornelof, E. (eds.) Critical Loads for Nitrogen – a Workshop Report. *Nord* 41: 307–353.

Ashby, E. 1936. Statistical ecology. *Botanical Review* 2(5): 214–219.

—— 1948. Statistical ecology 2. *Botanical Review* 14(4): 222–234.

Aubreville, A. 1971. *Regeneration patterns in the closed forest of Ivory Coast. World Vegetation Types.* Columbia University Press. New York. Pp. 41–55.

Battaglia, M. and Sands, P.J. 1998. Process-based forest productivity models and their application to forest management. *Forest Ecology and Management* 102: 13–32.

Bazilevich, N.I. 1978. An attempt of conceptual soil modelling. *Doklady Academy of Sciences of USSR* 240: 959–962 (In Russian with English summary).

Bazykin, A.D. 1985. Mathematical Biophysics of Interacting Populations. Nauka. Moscow. 181 p. (In Russian).

Bazzaz, F.A. and Sipe, T.W. 1987. *Physiological ecology, disturbance, and ecosystem recovery.* In: Schulze, E.-D. and Zwolfer, H. (eds.). Potential and limitations of ecosystem analysis. Springer Verlag. New York. Pp. 203–227.

Bazykin, A.D., Berezovskaya, F.S., Isaev, A.S. and Khlebopros, R.G. 1995. Analysis of stereotypes of forest insect dynamics. *Journal of General Biology, Moscow* 56 (2): 191–199. (In Russian with English summary).

Benjamin, L.R., Hardwick, R.C. 1986. Sources of variation and measures of variability in even-aged stands of plants. *Annals of Botany* 58: 757–778.

Benjamin, L.R. and Sutherland, R.A. 1992. *A comparison of models to simulate the competitive interactions between plants in even-aged monocultures.* In: DeAngelis, D.L. and Gross, L.S. (eds.) Individual-Based Models and Approaches in Ecology: Populations, Communities and Ecosystems. NY Academic Press. New York. Pp. 455–471.

Berezovskaya, F.S. and Karev, G.P. 1990. *Mosaic conception of spatio-temporal structure and tree stand dynamic modeling.* In: International symposium "North Forests". Goskomles. Moscow 3: 3–11 (In Russian).

Berezovskaya, F.S. and Karev, G.P. 1994. *Analytical approach to ecophysiological forest modelling. Computer reference-information system.* IAEA and UNESCO, International Centre for Theoretical Physics. Internal Report. 19 p.

Berezovskaya, F.S. and Karev, G.P. 1997. *New approaches to qualitative behaviour modelling of complex systems.* Proceedings of International Conference on Informatics and Control. St. Peterburg.

Berezovskaya, F.S., Karev, G.P. and Shvidenko, A.Z. 1991. *Modelling Stands' Dynamics: Eco-Physiological Approach.* Research and Information Center on Forest Resources. Moscow. 84 p. (In Russian).

Berezovskaya, F.S., Karev, G., Kisliuk, O., Khlebopros, R. and Tselniker, Yu. 1993. *Fractal approach to computer-analytical modeling of tree crown.* IC/92/267. Internal Report I.T.C.P.P. Triest.

Bertalanffy, L. 1942. *Theoretische Biologie.* Borntraeger. Berlin-Zenlendorf XVI 3025.

Besag, J. 1974. Spatial interaction and the statistical analysis of lattice systems. *Journal of Royal Statistical Society* B36: 192–236.

—— 1975. Statistical analysis of non-lattice data. *The Statistician* 24: 179–195.

Besag, J. and Gleaves, J. 1973. On the detection of spatial pattern in plant communities. *Bulletin of Institute International Statistics* 45:153–158.

Bevins, C.D., Andrews, P.L. and Keane, R.E. 1995. Forest succession modelling using the Loki software. *Forest Science, Prague* 41: 158–162

Bikhele, Z.I., Moldau, Kh.A. and Ross, Yu.K. 1980. *Mathematical Modelling of Plants' Transpiration and Photosynthesis under Soil Moisture Deficiency.* Hydrometeoizdat. Leningrad. 134 p. (In Russian).

Bogatyrev, B.G., Kirilenko, A.P. and Tarko, A.M. 1988. *Spatially Distributed Models of Biosphere.* Computer Center of Academy of Sciences. Moscow. (In Russian).

Bondarenko, O.N. and Liapunov, A.A. 1973. *On mathematical models of soil forming processes.* In: Cybernetic Approaches in Biology. Nauka. Novosibirsk. Pp. 210–231. (In Russian).

Bondarenko, N.F., E.E. Zhukovsky, I.G. Mushkin, S.V. Nerpin, R.A. Poluektov and I.B. Uskov. 1982. *Modelling of Agroecosystems' Productivity.* Hydrometeoizdat. Leningrad. 264 p. (In Russian).

Bosatta, E. and Ågren, G.I. 1991. Dynamics of carbon and nitrogen in the organic matter of the soil: a generic theory. *American Naturalist* 138: 227–245.

Bossel, H. 1991. Modelling forest dynamics: moving from description to explanation. *Forest Ecology and Management* 42: 129–142.

—— 1994. *TREEDYN3 Forest Simulation Model – mathematical model, program documentation, and simulation results.* Forschungszentrum Waldökosysteme der Universität Göttingen. Göttingen.

Bossel, H., Krieger, H., Schäfer, H. and Trost, N. 1991. Simulation of forest stand dynamics using real structure process models. *Forest Ecology and Management.* 42: 3–21.

Botkin, D.B. 1981. *Causality and succession.* In: West, D.C., Shugart, H.H. and Botkin, D.B. (eds.). Forest Succession – Concepts and Applications. Springer-Verlag. New York. Pp. 36–55. Botkin, D.B. 1993. *Forest Dynamics. An Ecological Model.* Oxford University Press. Oxford, New York. 310 p.

Botkin, D.B., Janak, J.F. and Wallis, J.R. 1972. Some ecological consequences of a computer model of forest growth. *Journal of Ecology* 60: 849–872.

Bouma, J. and Hack-Ten Broecke, M.J.D. 1993. Simulation modelling as a method to study land qualities and crop productivity related to soil structure differences. *Geoderma* 57: 51–67.

Breckling, B. and Müller, F. (eds.) 1994. State-of-the-Art in Ecological Modelling. Proceedings of ISEM's 8th International Conference, Kiel, 28 Sept.–2 Oct. 1992. *Ecological Modelling* 75. 683 p.

Bugmann, H.K.M. 1996. A simplified forest model to study species composition along climate gradient. *Ecology* 77: 2055–2074.

Bugmann, H. and Cramer, W. 1998. Improving the behaviour of forest gap models along drought gradients. *Forest Ecology and Management* 103: 247–263.

Busykin, A.I., Gavrikov, V.I., Sekretenko, O.P. and Chlebopros, R.M. 1985. *Analysis of Forest Coenoses Structure.* Nauka. Novosibirsk. 85 p. (In Russian).

Busykin, A.I., Sekretenko, O.P. and Chlebopros, R.M. 1987. *The structure of forest coenoses.* Lectures in Memory of V.I. Sukachev 5. Nauka. Moscow (In Russian).

Cannell, M.G.R., Rothery, P. and Ford, E.D. 1984. Analysis of the dynamics of competition within stands of *Picea sitchensis* and *Pinus contorta. Annals of Botany* 53: 349–362.

Cherkashin, A.K. 1981. *The model of spatial and age structure of forest.* In: Models of Forest Resource Management. Nauka. Pp. 231–245. (In Russian).

Chertov, O.G. 1981. *Ecology of Forest Lands.* Nauka. Leningrad. 192 p. (In Russian).

—— 1983a. Mathematical model of a single plant ecosystem. *Journal of General Biology, Moscow* 44: 406–414 (In Russian with English summary).

—— 1983b. Qualitative approach to the species' ecological parameters evaluation with special reference to Scots pine. *Botanical Journal, Leningrad* 68: 1318–1324 (In Russian with English summary).

—— 1985. Simulation model of forest litter and floor mineralization and humification). *Journal of General Biology, Moscow* 46: 794–804 (In Russian with English summary).

—— 1990. SPECOM – a single tree model of pine stand/raw humus soil ecosystem. *Ecological Modelling* 50: 107–132.

Chertov, O.G., Gladkov, E.G., Vladimirova, S.K. and Vladimirov, V.K. 1990. Prediction of environmental changes in Kara-Kala region of Turkmenian Republik. *Problems of Desert Development* 5: 60–64 (In Russian with English translation in USA).

Chertov, O.G. and Komarov, A.S. 1995a. *On the mathematical theory of soil forming processes. I. Theoretical background. II. SOMM – a model of soil organic matter dynamics, III. Basic ideas of mineral phase modelling.* Pushchino Research Centre of RAS. Pushchino. 39 p.

Chertov, O.G. and Komarov, A.S. 1995b. *Dynamic modelling of Scots pine, Norway spruce and silver birch ecosystems in European boreal forests.* EFI Project Record 519. 152 pp.

Chertov, O.G. and Komarov, A.S. 1996. *Model of soil organic matter and nitrogen dynamics in natural ecosystems.* In: D.S. Powlson, P. Smith and J. Smith, (eds.) Evaluation of Soil Organic Matter Models Using Existing Long-Term Datasets. NATO ASI Vol. I 38, Springer-Verlag, Heidelberg. Pp. 231–236.

Chertov, O.G. and Komarov, A.S. 1997a. SOMM: a model of soil organic matter dynamics. *Ecological Modelling* 94: 177–189.

Chertov, O.G. and Komarov, A.S. 1997b. *Simulation model of Scots pine, Norway spruce and silver birch ecosystems.* Proc. of IBFRA 96 Annual Conference, St. Petersburg, August 1996. In press.
Chertov, O.G., Komarov, A.S. and Tsiplianovsky, A.M. 1998a. A combined simulation model of Scots pine, Norway spruce and Silver birch ecosystems in European boreal zone. *Forest Ecology and Management.* In press.
Chertov, O.G., Komarov, A.S. and Tsiplianovsky, A.M. 1998b. Simulation of soil organic matter and nitrogen accumulation in Scots pine plantations on bare parent material using forest combined model EFIMOD. *Plant and Soil.* In press.
Chertov, O.G., Komarov, A.S., Tsiplianovsky, A.M., Bykhovets, S.S. 1998c. Simulation Model EFIMOD of the system "Mixed Stand/Soil" in European Boreal Zone. Manuscript to be submitted as EFI Working Paper. 95 p.
Chertov, O.G., Prokhorov, V.M. and Kvetnaya, O.M. 1978. On modelling soil processes. *Soviet Soil Science, Moscow* 11: 138–146. (In Russian with English summary).
Chertov, O.G. and Razumovsky, S.M. 1980. On ecological trends of soil forming processes. *Journal of General Biology, Moscow* 41: 386–396 (In Russian with English summary).
Chestnykh, O.V. 1986. *Modelling production process of tree layer in South taiga Norway spruce stand.* PhD Thesis, Moscow. 22 p. (In Russian).
Chetverikov, A.N. 1985. *Modelling forest biogeocoenoses.* In: Svirezhev Ju.M., (ed.) Mathematical Modelling of Biogeocoenotic Prosesses. Nauka. Moscow. Pp. 37–51 (In Russian).
Chumachenko, S.I. 1992. *Bioecological model of uneven-aged forest coenosis.* Doctoral Dissertation at Moscow State Pedagogical University. 129 p. (In Russian).
Chumachenko, S.I., Syssouev, V.V., Palyonova, M.M., Bredikhin, M.A. and Korotkov, V.N. 1996. *Simulation modelling of heterogeneous uneven-aged stands spatial dynamics taking into account silvicultural treatment.* IUFRO Conference. Copenhagen. Pp. 484–492.
Clements, F.E. 1916. *Plant Succession.* New York.
Cole, L.C. 1946. A theory for analyzing contagiously distributed populations. *Ecology* 27(4): 329–341.
Coates, K.D. and Burton, Ph.J. 1997. A gap-based approach for development of silvicultural systems to address ecosystem management objectives. *Forest Ecology and Management* 99: 337–354.
Coleman, K. and Jenkinson, D.S. 1995. *ROTHC-26.3: A Model for the Turnover of Carbon in Soil.* IACR Rothamsted. Harpenden, Herts.
Cox, D.R. and Isham, V. 1980. *Point Processes.* Chapman and Hall. London. 188 p.
Dale, V.H., Doyle, T.W. and Shugart, H.H. 1985. A comparison of tree growth models. *Ecological Modelling* 29: 145–170.
DeAngelis, D.L. and Gross, L.S. (eds.) 1991. *Individual-Based Models and Approaches in Ecology: Populations, Communities and Ecosystems.* Chapman and Hall. New York and London.
De Ruiter, P.C. and Van Faassen, H.G. 1994. A comparison between an organic matter dynamics model and a food web model simulating nitrogen mineralization in agro-ecosystems. *European Journal of Agronomy* 3: 347–354.
De Vasconcelos, M.J.P and Zeigler, B.P. 1993. Discrete-event simulation of forest landscape response to fire disturbances. *Ecological Modelling* 65: 177–198.
Diggle, P.J. 1976. A spatial stochastic model of interplant competition. *Journal of Applied Probability* 13: 662–671.
Diggle, P. 1983. *Statistical Analysis of Spatial Point Patterns.* Academic Press. London.
Duchaufour, Ph. 1961. *Precis de pedologie.* Paris. Russian translation Mir. Moscow 1970.
Dylis, N.V. 1969. *Structure of Forest Biocoenoses.* Nauka. Moscow. 55 p. (In Russian).
Eitingen, G.R. 1962. *The Selected Papers.* Selkhozgiz. Moscow (In Russian).
Ek, A.K. and Monserud, R.A. 1974. *FOREST: a computer model for simulating the growth*

and reproduction of mixed species forest. Research Report No 2634. School of Natural Resources. University of Wisconsin.
Feller, C. and Beare, M.H. 1997. Physical control of soil organic matter dynamics in the tropics. *Geoderma* 79: 69–116.
Fisher, R.A. and Miles, R.E. 1973. The role of spatial pattern in the competition between crop plants and weeds: A theoretical analysis. *Mathematical BioSciences* 18: 335–350.
Flechsig, M., Erhard, M. and Wenzel, V. 1994. Simulation-based regional models – concept, design and application. *Ecological Modelling* 75/76: 601–608.
Ford, E.D. 1975. Competition and stand structure in some even-aged plant monocultures. *Journal of Ecology* 63: 311–333.
Ford, E.D. and Diggle, P.J. 1981. Competition for light in a plant monoculture modelled as a spatial stochastic process. *Annals of Botany* 48: 481–500.
Ford, E.D. and Sorrensen, K.A. 1992. *Theory and models of inter-plant competition as a spatial process.* In: DeAngelis, D.L and Gross, L.J. (eds.) Individual-Based Models and Approaches in Ecology. Chapman and Hall. New York and London. Pp. 363–407.
Franko, U., Oelschlael, B. and Schenk, S. 1995. Simulation of temperature, water and nitrogen dynamics using the model CANDY. *Ecological Modelling* 81: 213–222.
Frelich, L.E. and Lorimer, C.G. 1991. A simulation of landscape-level stand dynamics in northern hardwood region. *Journal of Ecology* 79: 223–233.
Friend, A.D., Stevens, A.K., Knox, R.G. and Cannel, M.G.R. 1997. A process-based, terrestrial biosphere model of ecosystem dynamics (Hybrid v3.0). *Ecological Modelling* 95: 249–287.
Frost, H.J. and Thompson, C.V. 1987. The effect of nucleation conditions on the topology and geometry of two-dimensional grain structures. *Acta Metallurgia* 35: 529–540.
Galitsky, V.V. 1979. On collective self-oppression in a uniform plant community and oscillating changes of the biomass of its members. *Doklady of USSR Academy of Sciences* 246: 1013–1015. (In Russian with English summary).
—— 1982. *On modelling the plant community dynamics.* In: Trappl, R. (ed.) Cybernetics and Systems Research. North-Holland Publishing Company. Pp. 677–682.
Galitsky, V.V., Glotov, N.V., Komarov, A.S., Krylov, A.A. and Semerikov, L.F. 1982. *Modelling of the spatial pattern of genetic structure of the population of trees.* In: Evolutionary Genetics. Leningrad State University. Leningrad. Pp. 141–159. (In Russian).
Galitsky, V.V. and Komarov, A.S. 1974. *Non-free Growth of Organism's Biomass.* Biological Research Center of the USSR Academy of Sciences. Pushchino. (In Russian).
—— 1987. *Discrete models of plant populations.* In: Samarsky, A.A. (ed.) Mathematical Modelling. Nonlinear Systems. Nauka. Moscow. Pp. 59–103. (In Russian).
Gates, D.J. 1978. Bimodality in even-aged plant monocultures. *Journal of Theoretical Biology* 71: 525–540
—— 1980a. Competition between two types of plants located at random on a lattice. *Mathematical BioSciences* 48: 157–194.
—— 1980b. Competition between two types of plants with specified neighbour configurations. *Mathematical BioSciences* 48: 195–209.
—— 1982. Analysis of some equations of growth and competition in plantations. *Mathematical BioSciences* 59: 17–32.
Gates, D.J. and Westcott, M. 1978. Zone of influence models for competition in plantation. *Advances of Probability* 10: 499–537.
Gates, D.J., O'Connor, A.J. and Westcott, M. 1979. Partitioning the union of disks in plant competition models. *Proceedings of Royal Society London* A 367: 59–79.
Gatsuk, L.E., Smirnova, O.V., Vorontsova, L.I., Zaugolnova, L.B. and Zhukova, L.A. 1980. Age states of plants of various growth forms: a review. *Journal of Ecology* 68: 675–696.

Gavrikov, V.L. 1985. Age dynamics in tree populations. In: Analysis of structure of arboreal coenoses. Novosibirsk. Nauka. Pp. 50–70. (In Russian).

Gavrikov, V.L., Grabarnik, P.Ya. and Stoyan, D. 1993. Trank-top relations in a Siberian pine forest. *Biometrical Journal* 35: 487–498.

Gilbert, E.N. 1962. Random subdivisions of space into crystals. *Annals of Mathematical Statistics* 33: 958.

Gilderman, Yu.I., Kudrina, K.I. and Poletaev, I.A. 1970. *L-system models: systems with limiting factors.* In: Cybernetics' Investigations. Soviet Radio Publishers. Moscow. Pp. 165–210. (In Russian).

Gilmanov, T.G. 1974. Linear model of long-term dynamics of soil organic matter. *Herald of Moscow State University, Biol., Soil Sci.,* No 6: 116–123. (In Russian with English summary).

Gimmelfarb, A.A., Ginzburg, L.R., Poluektov, R.A., Pikh, Yu.A., and Ratner, V.A. 1974. *Dynamics of Biological Populations.* Nauka. Moscow. 456 p. (In Russian).

Goldberg, D.E. 1987. Neighborhood components in an old-field plant community. *Ecology* 68: 1211–1223.

—— 1990. *Components of resource competition in plant communities.* In: Grace, J.B. and Tilman, D. (eds). Perspectives on Competition. Academic Press. San Diego, California. Pp. 27–49.

Gonchar-Zaikin, P.P. and Zhuravlev, O.S. 1979. *Simple model of soil humus dynamics.* In: Theoretical Bases and Quantitative Methods of Crop Programming. Agro-Physical Institute. Leningrad. Pp. 156–165 (In Russian).

Goto, N., Sakoda, A. and Suzuki, M. 1994. Modelling of soil carbon dynamics as a part of carbon cycle in terrestrial ecosystems. *Ecological Modelling* 74: 183–204.

Goulard, M., Sarkka, A. and Grabarnik, P. 1996. Parameter estimation for marked Gibbs point processes through the maximum pseudo-likelihood method. *Scandinavian Journal of Statistics* 23: 353–364.

Grabarnik, P.Ya. and Komarov, A.S. 1980. *Statistical analysis of spatial structures. Methods using distances between points.* Pushchino. 48 p. (In Russian).

Grabarnik, P.Ya. and Komarov, A.S. 1981. *Spatial structure of Scots pine stands.* In: Molchanov, A.M. (ed.) Modeling of Biogeocoenotical Processes. Nauka. Moscow. Pp. 59–81. (In Russian).

Grabarnik, P.Ya., Komarov, A.S., Nosova, L.M. and Radin, A.I. 1992. *Analysis of stand's spatial structure: correlation measures using.* In: Mathematical Modelling of Plant Populations and Phytocenoses. Nauka. Moscow. Pp. 74–85. (In Russian).

Grant, R.F., Juma, N.G. and McGill, W.B. 1993a. Simulation of carbon and nitrogen transformations in soil: mineralization. *Soil Biology and Biochemistry* 25: 1317–1329.

Grant, R.F., Juma, N.G. and McGill, W.B. 1993b. Simulation of carbon and nitrogen transformations in soil: microbial biomass and metabolic products. *Soil Biology and Biochemistry* 25: 1331–1338.

Greig-Smith, P. 1967. *Quantitative Plant Ecology.* Nauka. Moscow. 316 p. (Russian translation).

Gulpin, M.E. and Hanski, I. (eds.). 1991. *Metapopulation Dynamics.* Academic Press. London.

Gurtin, M. and MacCamy, R. 1977. Population dynamics with age-dependence. *Nonlinear Analysis and Mechanics* 3: 1–35.

Gurtsev, A.I. and Tselniker, Ju.L. 1997. Fractal structure of a branch. *Russian Forest Science, Moscow.*

Gussakov, S.V. and Fradkin, A.I. 1990. *Computer Simulation of the Spatial Structure of the Forest Biocoenoses.* Navuka i Technika. Minsk. 112 p. (In Russian).

Gussakov, S.V. and Fradkin, A.I. 1992. *Methods of the modelling space structure of forest phytocoenoses.* In: Logofet, D.O. (ed.) Mathematical Modelling of Plant Populations and Phytocenoses. Nauka. Moscow. Pp. 91–105. (In Russian).

Gyllenberg, M., Osipov, A.V. and Soderback, G. 1996. Bifurcation analysis of a metapopulation model with sources and sinks. *J. Nonlinear Science* 6: 329–366.

Haase, P. 1995. Spatial pattern analysis in ecology based on Ripley's K-function: Introduction and methods of edge correction. *Journal of Vegetation Science*: 575–582.
Habets, A.S.J. 1991. *FARM, a more objective calculating model for arable-, diary-, beef- and mixed farms*. Thesis. Dept. of Ecological Agriculture, Wageningen Agricultural University. Wageningen.
Hansen, S., Jensen, H.E., Nielsen, N.E. and Svendsen, H. 1991. Simulation of nitrogen dynamics and biomass production in winter wheat using the Danish simulation model DAISY. *Fertilizer Research* 27: 245–259.
Hansky, I., Moilanen, A. and Gyllenberg, M. 1996. Minimum viable metapopulation size. *The American Naturalist*, 147 (4): 527–541.
Hari, P. and Kellomäki, S. 1981. Modelling of the functioning of a tree in a stand. *Studia Forestalia Suecica* 160: 39–42.
Harper, J.L. 1977. *Population Biology of Plants*. Academic Press Inc. New York.
Hassink, J. 1994. Active organic matter fractions and microbial biomass as predictors of N mineralization. *European Journal of Agronomy* 3: 257–265.
Hines, W.G.S. and O'Hara Hines, R.J. 1965. The Eberhardt statistics and detection of nonrandomness of spatial point distributions. *Biometrika* 52(3): 345–353.
Hopkins, B. and Skellam, T.G. 1954. A new method for determining the type of distribution of plant individuals. *Annals of Botany* 18(70): 213–227.
Horn, H.S. 1975. *Markovian property of forest succession*. In: M.L.Cody and J.M.Diamond (eds.). Ecology and Evolution of Communities. Harvard University Press. Cambridge. Pp. 196–211.
Hugget, R.J. 1975. Soil landscape systems: a model of soil genesis. *Geoderma* 13: 1–22.
Hunt, H.W. 1977. A simulation model for decomposition in grasslands. *Ecology* 58: 469–484.
Hunt, H.W., Coleman, D.C., Ingham, E.R., Ingham, R.E., Elliot, E.T., Moore, J.C., Rose, S.L., Reid, C.P.P. and Morley, C.R. 1987. The detrital food web in a shortgrass prairie. *Biology and Fertility of Soils* 3: 57–68.
Huston, M. and DeAngelis, D. 1987. Size bimodality in monospecific populations: a critical review of potential mechanisms. *American Naturalist* 129: 678–707.
Huston, M., DeAngelis, D. and Post, W. 1988. New computer models unify ecological theory. *BioScience* 38 (10): 682–691.
Huxley, J.S. 1932. *Problems of Relative Growth*. Diul Press. New York. 296 p.
Hynynen, J. 1995. *Modelling tree growth for managed stands*. The Finnish Forest Research Institute. Research Paper 576. 59 p.
Ingestad, T. and Ågren, G.I. 1991. The influence of plant nutrition on biomass allocation. *Ecological Applications* 1: 168–174.
Isaev, A.S., Khlebopros, R.G., Nedoresov, L.V., Kondakov, Ju.P.P. and Kiselev, V.V. 1984. *Dynamics of Number of Forest Insects*. Nauka. Novosibirsk. (In Russian).
Jenkinson, D.S. 1990. *The turnover of organic carbon and nitrogen in soil*. Philosophical Transactions of Royal Society, London B 329. Pp. 361–369
Jenkinson, D.S., Hart, P.B.S., Rayner, J.H. and Parry, L.C. 1987. *Modelling the turnover of organic matter in long-term experiments at Rothamsted*. INTECOL Bulletin 15. Pp. 1–8.
Jenkinson, D.S. and Rayner, J.H. 1977. The turnover of soil organic matter in some of the Rothamsted classical experiments. *Soil Science* 123: 298–305.
Jenny, H., Gessel, S.P. and Bingham, F.T. 1949. Comparative study of decomposition rates of organic matter in temperate and tropical regions. *Soil Science* 69: 419–432.
Johnson, W.A. and Mehl, R.F. 1939. Reaction kinetics in processes of nucleation and growth. *Transactions of American Institute Mining and Engineering* 135: 416–458.
Karev, G.P. 1984. Mathematical growth model of multi-layer tree stands. *Proceedings of Siberian Branch of USSR Academy of Sciences, ser. Biology* 23(2): 8–14 (In Russian with English summary).

—— 1985a. On modelling multi-species tree stands. In: Problems of Ecological Monitoring and Ecosystems Modelling. *Hydrometeoizdat, Leningrad* 7: 227–233 (In Russian with English summary).
—— 1985b. Mathematical model of light competition in light-limiting self-thinning tree stands. *Journal of General Biology, Moscow* 46: 75–90 (In Russian with English summary).
—— 1985c. *On one approach to modelling tree stand growth.* In: Mathematical Biophysics. Krasnoyarsk State University. Krasnoyarsk. Pp. 161–167 (In Russian).
—— 1992. Age-dependent population dynamics with several interior variables and spatial spread. *Ecological Modelling* 70: 277–288.
—— 1993. *Structural models and dynamics of tree populations.* Doctoral Dissertation. Center of Forest Ecology and Productivity. Moscow. 225 p. (In Russian).
—— 1994. *Structural models of phytocenosis succession dynamics and the problem of global climate change.* In: Global and Regional Ecological Problems. Krasnoyarsk. Pp. 57–69.
—— 1995a. *Models of trees' one generation population and cenons formation.* In: Problems of Monitoring and Forest Ecosystem Modelling. Ecos-Inform Publ. Moscow. Pp. 228–243. (In Russian).
—— 1995b. *Dynamics of forest ecosystem as a cenon metapopulation.* In: Problems of Monitoring and Forest Ecosystem Modelling. Ecos-Inform Publ. Moscow. Pp. 201–219. (In Russian).
—— 1996. Structural models of the dynamics of biological communities. *Doklady Russian Academy of Sciences, Mathematics* 54: 749–751.
—— 1997. On ergodic hypothesis in biocenology. *Doklady Russian Academy of Sciences, Mathematics* (in press).
Karev, G.P. and Treskov, S.A. 1982. Mathematical models of boundary effects in phytocenoses. In: Problems of Ecological Monitoring and Ecosystems Modelling. *Hydrometeoizdat, Leningrad* 5: 229–242. (In Russian).
Karev, G.P., and Skomorovski, Ju.L. 1998. Modelling of dynamics of single species tree stands. *Syberian Journal of Ecology.* (In Russian) (in press).
Karmanova, I.V., Ganina, N.V. and Dmitriev, V.V. 1992. *Mathematical models of spatial structure of forest community.* In: Mathematical Modelling of Plant Populations and Phytocenoses. Nauka. Moscow. Pp. 63–74. (In Russian).
Karpachevsky, L.O. 1981. *Forest and Forest Soil.* Nauka. Moscow. 264 p. (In Russian).
Kazimirov, N.I. and Morozova, R.M. 1973. *Biological Cycle in Karelian Spruce Forests.* Nauka. Leningrad. 176 p. (In Russian).
Kazimirov, N.I., Volkov, A.D., Ziabchenko, S.S., Ivanchikov, A.A. and Morozova, R.M. 1977. *Cycle of Matter and Energy in Scots Pine Forests of European North.* Nauka. Leningrad. 304 p. (In Russian).
Kazimirov, N.I. and Mitrukov, A.E. 1978. *Variability and mathematical model of Scots pine trees and stands phytomass.* In: Kazimirov, N.I. (ed.) Formation and Productivity of Scots Pine Stands in Karelian Republic and Murmansk district. Petrozavodsk. Pp. 142–148 (In Russian).
Kazimirov, N.I., Morozova, R.M. and Kulikova, V.K. 1978. *Organic Matter Pools and Flows in Pendula Birch stands of Middle Taiga.* Nauka. Leningrad. 216 p. (In Russian).
Kazimirov, N.I., Svirezhev, Ju.M., Tarko, A.M. and Chetverikov, A.N. (eds.) 1985. *Mathematical Modelling in Biogeocoenology.* Forest Research Institute. Petrozavodsk. 224 p. (In Russian).
Keane, R.E., Arno, S.F. and Brown, J.K. 1989. FIRESUM – An Ecological Process Model for Fire Succession in Western Conifer Forests. *USDA Forest Service Intermountain Research Station General Technical Report* INT-266.
Kellomäki, S., Väisänen, H., Hänninen, H., Kolström, T., Lauhanen, R., Mattila, U. and Pajari, B. 1992. SIMA: a model for forest succession based on the carbon and nitrogen cycles with application to silvicultural management of the forest ecosystem. *Silva Carelica* 22. 85 p.

Kellomäki, S., Väisänen, H. and Strandman, H. 1993. *FinnFor: a model for calculating the response of the boreal forest ecosystem to climate changes.* Research Note No. 6. Faculty of Forestry, University of Joensuu. Joensuu, Finland. 120 p.

Kenkel, N.L. 1988. Pattern of self-thinning in Jack pine: testing the random mortality hypotheses. *Ecology* 64: 1017–1024.

Kent, B.M. and Dress, P. 1979. On the convergence of forest stand spatial pattern over time: The case of random initial spatial pattern. *Forest Science* 25: 445–451.

—— 1980. On the convergence of forest stand spatial pattern over time: The case of regular and aggregated initial spatial pattern. *Forest Science* 26: 10–22.

Khilmi, G.F. 1957. *Theoretical Forest Biogeophysics.* Hydrometeoizdat. Leningrad. 296 p. (In Russian).

—— 1966. *Foundations of Biosphere Physics.* Hydrometeoizdat. Leningrad. 300 p. (In Russian).

—— 1976. *Energy and Productivity of Terrestrial Vegetation Cover.* Hydrometeoizdat. Leningrad. 62 p. (In Russian).

Kienast, F. 1987. *FORECE – a forest succession model for southern central Europe.* Oak Ridge National Laboratory. Oak Ridge, Tennessee. ORNL/TM–10575.

Kimmins, J.P. 1990. Modelling the sustainability of forest production and yield for a changing and uncertain future. *Forestry Chronicle* 66: 271–280.

Kimmins, J.P. and Scoullar, K.A. 1979. *FORCYTE (Forest Cycling Trend Evaluator): a computer simulation model to examine long-term consequences for site nutrient capital and productivity of intence forest harvesting.* Report. Department of Environment, Canadian Forestry Service. Pettawawa. 90 p.

Kirkpatrick, S., Gelett, C.D. and Vecchi, M.P. 1983. Optimization by simulated annealing. *Science* 220: 621–680.

Kirschbaum, M.U.F., King, D.A., Comins, H.N., et al. 1994. Modelling forest response to increasing CO_2 concentration under nutrient-limited conditions. *Plant, Cell and Environment* 17: 1081–1099.

Kiviste, A.K. 1988. *Functions of Forest Growth.* Tartu (In Russian).

Kline, J.R. 1973. Mathematical simulation of soil-plant relationships and soil genesis. *Soil Science* 115: 240–249.

Knyazkov, V.V., Logofet, D.O. and Tursunov, R.D. 1992. *Non-homogeneous Markov model for plant succession in Tigrovaya Balka State Reserve.* In: Logofet, D.O. (ed.) Mathematical Modelling of Plant Populations and Phytocoenoses. Nauka. Moscow. Pp. 37–49. (In Russian).

Kofman, G.V. 1986. *Growth and Form of Trees.* Nauka. Novosibirsk. 211 p.(In Russian).

Kolmogoroff, A.N. 1937. Statistical theory of crystallization of metals. *Bulletin of Academy of Sciences of the USSR. Mathematics Seria* 1: 355–359 (In Russian and German).

Komarov, A.S. 1979. *Markov fields and plant communities.* In: Dobrushin, R.L., Kryukov, V.I. and Toom, A.L. (eds.) Interactive Markov processes and their applications to multicomponent systems. Pushchino. Pp. 7–21. (In Russian).

Komarov, A.S. 1988. *Mathematical models in population biology of plants.* In: L.B. Zaugolnova et al. (eds.) Coenopopulations of Plants. Nauka. Moscow. Pp. 137–155 (In Russian).

Komarov, A.S. and Grabarnik, P.Ya. 1980. *Statistical Analysis of Spatial Patterns. Methods Using Distances Between Nearest Neighbours.* Pushchino Scientific Centre of Russian Academy of Sciences. Pushchino. 29 p. (In Russian).

Korzukhin, M.D. 1980. Age dynamics of population of trees being strong edificators. In: Problems of Ecological Monitoring and Ecosystems Modelling. *Hydrometeoizdat, Leningrad* 3: 162–178. (In Russian).

—— 1985. Some strategies of plant ontogenesis under competition. Problems of Ecological Monitoring. *Hydrometeoizdat, Leningrad* 7: 234–241 (In Russian with English summary).

Korzukhin, M.D., Matskiavichus, V.K. and Antonovsky, M.Ya. 1989. Periodical behavior of age-distributed tree population. Problems of Ecological Monitoring. *Hydrometeoizdat, Leningrad* 12: 284–310 (In Russian with English summary).

Korzukhin, M.D. and Semevsky, F.N. 1992. *Forest Synecology*. Hydrometeoizdat. St.Petersburg. 192 p. (In Russian with English summary).

Korzukhin, M.D. and Ter-Mikaelian, M.G. 1987. Optimal plant ontogenesis with special reference to its protection and competition. Problems of Ecological Monitoring. *Hydrometeoizdat, Leningrad 10: 244–256* (In Russian with English summary).

—— 1995. An individual tree-based model of competition for light. *Ecological Modelling* 79: 221–229.

Kostychev, P.A. 1889. *Formation and Properties of Humus*. St. Peterburg. Citation: 1951. Selected Works. Academy of Sciences Publ. Leningrad. Pp. 251–296 (In Russian).

Krieger, H., Schäfer, H. and Bossel, H. 1990. SPRUCOM – a simulation model of spruce stand dynamics under varying emission exposure. *System Analysis Modelling and Simulation* 7: 117–129.

Kudrina, K.N. 1973. *Mathematical model of higher plant*. In: Physiology of Plants' Adaptation to Soil Conditions. Nauka. Novosibirsk. Pp. 25–38 (In Russian).

Kull, K. and Kull, O. 1989. *Dynamic Modelling of Tree Growth*. Valgus. Tallinn. 232 p. (In Russian with English summary).

Kull, K. and Oja, T. 1984. Structure of physiological models of tree growth. *Proceedings of Academy of Sciences of the USSR, Biology* 33: 33–41. (In Russian with English summary).

Kuzmichev, V.V. 1977. *Pecularities of Stands Growth*. Nauka. Novosibirsk. (In Russian).

Kuznetsov, Ju.A., Antonovsky, M.Ya., Biktashev, V.N. and Aponina, E.A. 1996. Cross-diffusion model of forest boundary dynamics. In: Problems of Ecological Monitoring. *Hydrometeoizdat, St. Petersburg* 14: 213–226. (In Russian with English summary).

Lavelle, P. 1987. Interactions, hierarchies et regulations dans le sols a la recherche d'une nouvelle approache conceptuelle. *Review Ecologie et Biologie du Sol* 24: 219–229.

Leemans, R. 1992. Simulation and future projection of succession in a Swedish broad-leaved forest. *Forest Ecology and Management* 48: 305–319.

Leskov, A.I. 1927. About variance of distances between trees. *Proceedings of Leningrad Naturalists Society* 1 (In Russian).

Levin, S.A., Hallam, T.G. and Gross, L.J. (eds.) 1989. *Applied Mathematical Ecology*. Springer Verlag.

Levine, E.R., Ranson, K.J., Smith, J.A., Williams, D.L., Knox, R.G., Shugart, H.H., Urban, D.L. and Lawrence, W.T. 1993. Forest ecosystem dynamics: linking forest succession, soil process and radiation models. *Ecological Modelling* 65: 199–219.

Li, C., Folkring, S. and Harris, R. 1994. Modelling carbon biogeochemistry in agricultural soils. *Global Biochemical Cycles* 8: 237–254.

Liddle, M.J., Budd, C.S.J. and Hutchings, M.J. 1982. Population dynamics and neighborhood effects in establishing swards of *Festuca rubra*. *Oikos* 38: 52–59.

Liebig, J. 1876. *Chemistry Applications to Farming and Physiology*. Russian translation by OGIZ-Selkhozgiz Moscow and Leningrad, 1936.

Lindner, M., Bugmann, H., Lasch, P., Flechsig, M. and Cramer, W. 1997. Regional impacts of climatic change of forests in the state of Brandenburg, Germany. *Agricultural and Forest Meteorology* 84: 123–135.

Liu, J. and Ashton, P.S. 1995. Individual-based simulation models for forest succession and management. *Forest Ecology and Management* 73: 157–175.

Lotka, A. 1925. *Elements of Physical Biology*. Williams and Wilkins. 450 p.

Madgwick, H.A.I., Olah, F.D. and Burkhart, H.E. 1977. Biomass of open grown Virginia pine. *Forest Science* 23: 89–91.

Mahin, K.W., Hanson, K. and Morris, J.W. 1980. Comparative analysis of the cellular and Johnson-Mehl microstructures through computer simulation. *Acta Metallurgica* 28: 443–453.

Mäkelä, A. 1990. *Modeling structural-functional relationships in whole-tree growth: resource allocation.* In: Dixon, R.K., Mehldahl, R.S., Ruark, G.A. and Warren, W.C. (eds.) Process Modelling of Forest Growth Responses to Environmental Stress. Timber Press. Portland. Pp. 81–95.
Mandelbrot, B.B. 1977. *The Fractal Theory of Nature.* Freedman and Co. New York. 468 p.
Martynov, A.N. 1976. A dependency between Scots pine biometrical characteristics and available square of nutrition. *Soviet Forest Science, Moscow,* 5: 85–88 (In Russian with English summary).
McMurtrie, R. 1981. Suppression and dominance of trees with overlapping crowns. *Journal of Theoretical Biology* 89: 151–174.
Mead, R. 1966. A relationship between individual plant spacing and yield. *Annals of Botany* 30: 301–309.
Mejering, J.L. 1953. Interface area, edge length and number of dertices in crystal aggregates with random nucleation. *Philips Research Reports* 8: 270–290.
Metz, J.A.J. and Deikmann, O. (eds.) 1986. The Dynamics of Physiologically Structured Populations. *Lecture Notes in Biomathematics* 68. Springer Verlag.
Miina, J., Kolström, T. and Pukkala, T. 1991. An application of a spatial growth model of Scots pine on drained peatland. *Forest Ecology and Management* 41: 265–277.
Miles, R.E. 1969. A wide class of distributions in geometric probabilities. *Advances in Applied Probability* 2: 129–137.
—— 1972. The random division of space. *Supplement to Advances in Applied Probability: 243–266.*
Miroshnichenko, G.V. 1955. Theory and investigations of the intraspecies competition of plants. *Botanical Journal, Leningrad* 40: 408–410. (In Russian with English summary).
Mithen, R., Harper, J.L. and Weiner, J. 1984. Growth and mortality of individual plants as a function of "available area". *Oecologia* 62: 57–60.
Moeur, M. 1985. *COVER: A User's Guide to the CANOPY and SHRUBS Extension of the Stand Prognosis Model.* General Technical Report INT-190. Ogden, UT: U.S. Department of Agriculture, Forest Service, Intermountain Research Station. 49 p.
Mohren, G.M.J. 1994. *Modelling Norway spruce growth in relation to site conditions and atmospheric* CO_2. In: Veroustraite, F. et al. (eds.) Vegetation, Modelling and Climatic Change Effects. SPB Academic Publishing. The Hague. Pp. 7–22.
Mohren, G.M.J. and Kienast, F. (eds.) 1991. Modelling Forest Succession in Europe. *Forest Ecology and Management* 42: 1–143.
Mohren, G.M.J., van Hees, A.F.M. and Bartelink, H.H. 1991. Successional models as an aid for forest management in mixed stands in the Netherlands. *Forest Ecology and Management* 41: 111–127.
Molchanov, A.M. 1975. *Ecology and ergodicity.* In: Kovda, V.A. (ed.) Simulation Modelling and Ecology. Nauka. Moscow. Pp. 49–51 (In Russian).
—— 1985. *On Khinchin theorem.* Citation: Molchanov, A.M. (ed.). 1992. Nonlinearity Biology. Russian Academy of Sciences. Puschino. (In Russian).
Molina, J.A.E., Clapp, C.E., Shaffer, M.J., Chichester, F.W. and Larson, W.E. 1983. NCSOIL, a model of nitrogen and carbon transformations in soil: description, calibration and behavior. *Soil Science Society America Journal* 47: 85–91.
Moller, J. 1992. Random Johnson-Mehl tesselations. *Advances in Applied Probability* 24: 814–844.
—— 1994. *Lectures on Random Voronoi Tesselations.* Lecture Notes in Statistics. Springer-Verlag. New York.
—— 1995. Generation of Johnson-Mehl crystals and comparative analysis of models for random nucleation. *Advances in Applied Probability* 27: 367–383.
Monsi, M. and Saeki, T. 1953. Uber den Lichtfaktor in den Pflanzengesell-schaften und seine Bedeutung fur die Stoffproduktion. *Japan Journal of Botany* 14: 22–52.

Moran, M.A., Legovic, T., Benner, R. and Hodson, R.E. 1988. Carbon flow from lignocellulose: a simulation analysis of a detritus-based ecosystem. *Ecology* 69(5): 1525–1536.
Morozov, A.I. and Samoilova, E.M., 1993. On a problem of mathematical modelling of humus dynamics. *Euroasian Journal of Soil Science* 6: 24–32.
Morris, D.M., Kimmins, J.P. and Duckert, D.R. 1997. The use of soil organic matter as a criterion of the relative sustainability of forest management alternatives: a modelling approach using FORECAST. *Forest Ecology and Management* 94: 61–78.
Moshkalev, A.G. et al. 1984. *Reference Book on Forest Mensuration In Russian North-West.* Forest Technical Academy. Leningrad. 319 pp. (In Russian).
Mou, P., Mitchell, R.J. and Jones, R.H. 1993. Ecological field theory model: a mechanistic approach to simulate plant-plant interactions in southeastern forest ecosystems. *Canadian Journal of Forest Research* 23: 2180–2193.
Nakane, K. 1978. A mathematical model of the behavior and vertical distribution of organic carbon in forest soils. II. A revised model taking the supply of root litter into consideration. *Japan Journal of Ecology* 28: 169–177.
Newnham, R.M. 1968. *The generation of artificial populations of points (spatial pattern) on a plane.* Information Report. Forest Management Institute FMR–X–10. Ottawa. 19 p.
Neyman, J. 1939. On a new class of contagious distributions applicable in entomology and bacteriology. *Annals of Mathematical Statistics* 10(2): 147–154.
Nicolardot, B., Molina, J.A.E. and Allard, M.R. 1994. C and N fluxes between pools of soil organic matter: model calibration with long-term incubation data. *Soil Biology and Biochemistry* 26: 235–243.
Odum, E.P. 1971. *Fundamentals of Ecology.* Saunders. Philadelphia. 574 p. Russian translation 1975. Mir. Moscow.
Oja, T. 1985a. *Models of Stand Development.* Estonian Academy of Sciences. Tartu. (In Russian).
—— 1985b. A simple adaptive plant model. I. Model description. *Proceedings of Estonian Academy of Sciences, Biology* 34: 289–294 (In Russian with Estonian and English summaries).
—— 1986. A simple adaptive plant model. II. Model analysis. *Proceedings of Estonian Academy of Sciences, Biology* 35: 16–29 (In Russian with Estonian and English summaries).
—— 1989. *Modelling of succession [temporal continuum] of tree stand.* In: Paal, J.L., Oja, T.A. and Kolodyazhnyi, S.F. Taxonomic and Temporal Continuum of Plant Cover. Valgus. Tallinn. Pp. 119–148. (In Russian).
Oikawa, T. and Saeki, T. 1972. *Report 1971. Japan IBP/PP-Photosynthesis.* Level III Group. Tokyo. Pp. 107–116.
Okabe, A., Boots, B. and Sugihara, K. 1992. *Spatial Tesselations. Concepts and Applications of Voronoi Diagrams.* Wiley. Chichester.
Olson, J.S. 1963. Energy storage and balance of producers and decomposers in ecological systems. *Ecology* 44: 322–331.
Olson, R.L., Jr., Sharpe, P.J.H. and Wu, H. 1985. Whole-plant modelling: a continuous-time Markov (CTM) approach. *Ecological Modelling* 29: 171–187.
Päivinen, R. and Nabuurs, J.-G. 1996. *Large Scale Forestry Scenario Models – a Compilation and Review.* European Forest Institute Working Paper 10. European Forest Institute. Joensuu. 174 p.
Päivinen, R., Roihuvuo, L. and Siitonen, M. (eds.) 1996. *Large-Scale Forestry Scenario Models: Experiences and Requirements.* EFI Proceedings No. 5. European Forest Institute. Joensuu.
Pacala, S.W., Canham, C.D. and Silander, Jr., J.A. 1993. Forest models defined by field measurements: I. The design of a northeastern forest simulator. *Canadian Journal of Forest Research* 23: 1980–1988.
Pacala, S.W. and Silander, J.A. 1985. Neighborhood models of plant population

dynamics. I. Single species models of annuals. *American Naturalist* 125: 385–411.

Pacala, S.W. and Silander, J.A. Jr. 1987. Neighbourhood interference among velvet leaf, *Abutilon theofrasti*, and pigweed, *Amarathus retroflexus*. *Oikos* 48: 217–224. Springer-Verlag, Berlin, Heidelberg, 133–142

Parton, W.J. 1996. *Ecosystem model comparison: science of fantasy world?* In: Powlson, D.S., Smith, P. and Smith, J. (eds.). Evaluation of Soil Organic Matter Models. NATO ASI Series, I 38. Springer Verlag. Berlin Heidelberg.

Parton, W.J., Stewart, J.W.B. and Cole, C.V., 1988. Dynamics of C, N, P and S in grasslands soils: a model. *Biogeochemistry* 5: 109–131.

Parton, W.J., Scurlock, J.M.O., Ojima, D.S., Gilmanov., T.G., Scholes, R.J., Schimel, D.S., Kirchner, T., Menaut, J.-C., Seastedtd, T., Garcia Moya, E., Apinan Kamnalrut, and Kinyamario, J.I. 1993. Observations and modelling of biomass and soil organic matter dynamics for the grassland biome worldwide. *Global Biochemical Cycles* 7: 785–809.

Pastor, J. and Post, W.M. 1985. *Development of a Linked Forest Productivity – Soil Process Model.* Oak Ridge National Laboratory ORNL/TM-9519. 168 p.

Paustian, K., Parton, W.J. and Persson, J. 1992. Modelling soil organic matter in organic-amended and nitrogen-fertilized long-term plots. *Soil Science Society America Journal* 56: 476–488.

Paustian, K., Levine, E., Post, W.M., Ryzhova, I.M. 1997. The use of models to integrate information and understanding of soil C at the regional scale. *Geoderma* 79: 227–260.

Pegov, L.A., 1987. *Simulation modelling of horizontal structure in pendula birch stands.* PhD Thesis. Forest Technical Academy. Leningrad. 20 p. (In Russian).

Plentinger, M.C. and Penning de Vries, F.W.T. 1996. *CAMASE Register of Agro-Ecosystem Models.* AB-DLO, Wageningen.

Poletaev, I.A. 1966. On mathematical models of elementary processes in biogeocoenosis. *Problems of Cibernetics* 16: 171–190 (In Russian).

—— 1973. *On mathematical models of growth.* In: Physiology of Adaptation to Environmental Conditions. Nauka. Novosibirsk. Pp. 7–24.(In Russian).

—— 1975. *Use of the Liebig's principle in mathematical models of metabolic systems.* In V.A. Kovda, (ed.) Simulation Modelling and Ecology, Nauka. Moscow. Pp. 52–59 (In Russian).

Popadyuk, R.V. et al. 1994. *East-European broad-leaved forests.* Nauka. Moscow. 364 p. (In Russian).

Popadyuk, R.V. and Chumachenko S.I. 1991. Bioecological simulation model of many-species uneven-aged tree stand development. *Biological Sciences, Moscow State University* 8: 67–77. (In Russian with English summary).

Popov, A.I. and Chertov, O.G. 1993. On trophic function of soil organic matter. *Herald of St. Petersburg University, Seria* 3 *Biology* 3: 100–109 (In Russian with English summary).

Popov, A.I. and Chertov, O.G. 1996. The biogeocoenotic role of soil organic matter. *Herald of St. Petersburg University, Seria* 3 *Biology* 2: 88–97 (In Russian with English summary).

Post, W.M., King, A.W. and Wullschleger, S.D. 1996. *Soil organic matter models and global estimates of soil organic carbon.* In: Powlson, D.S., Smith, P. and Smith, J. (eds.). Evaluation of Soil Organic Matter Models. NATO ASI Series, Vol. I 38. Springer Verlag. Berlin, Heidelberg. Pp. 201–222.

Post, W.M. and Pastor, J. 1990. *An individual-based forest ecosystem model for projecting forest response to nutrient cycling and climate changes.* In: L.C. Wensel and G.S. Biging, (eds.) Forest Simulation Systems. University of California, Div. of Agriculture and Natural Resources. Bulletin 1927: 61–74.

Potter, C.S., Randerson, J.T., Field, C.B., Matson, P.A., Vitousek, P.M., Mooney, H.A. and Klooster, S.A. 1993. Terrestrial ecosystem production: a process model based on satellite and surface data. *Global Biogeochemical Cycles* 7: 811–841.

Powlson, D.S., Smith, P. and Smith, J. (eds.) 1996. *Evaluation of Soil Organic Matter Models*. NATO ASI Series, I 38. Springer Verlag. Berlin Heidelberg.

Prentice, I.C. and Helmisaari, H. 1991. Silvics of north European trees: compilation, comparisons and implications for forest succession modelling. *Forest Ecology and Management* 42: 79–93.

Prentice, I.C. and Leemans, R. 1990. Pattern and process and the dynamics of forest structure: a simulation approach. *Journal of Ecology* 78: 340–355.

Prusinkiewicz, Z. 1977. Application of the mathematical model of organic matter accumulation and decomposition for comparative study of various forest floor types. *Ekologia Polonica* 26: 343–357.

Pukkala, T. 1988. Effect of spatial distribution of trees on the volume increment of a young Scots pine stand. *Silva Fennica* 22: 1–17.

—— 1989a. Methods to describe the competition process in a tree stand. *Scandinavian Journal of Forest Research* 4: 187–202.

—— 1989b. Predicting diameter growth in even-aged Scots pine stands with a spatial and non-spatial model. *Silva Fennica* 23: 83–99.

Pukkala, T. and Kolström, T. 1987. Competition indices and the prediction of radial growth in Scots pine. 21: 55–76.

Pukkala, T. and Kolström, T. 1991. Effect of spatial pattern of trees on the growth of a Norway spruce stand. A simulation model. *Silva Fennica* 25: 117–131.

Pukkala, T., Kolström, T. and Miina, J. 1994. A method for predicting tree dimensions in Scots pine and Norway spruce stands. *Forest Ecology and Management* 65: 123–124.

Rabotnov, T.A. 1950. Life cycle of perennial grass plants in meadows coenose. *Proceedings of Komarov Botanical Institute. Seria* 3, *Geobotany* 6: 7–204 (In Russian with English summary).

—— 1978. *Phytocoenology*. Moscow State University Publ. Moscow. 383 p. (In Russian).

Rachko, P. 1978. Simulation model of tree growth. *Journal of General Biology, Moscow* 39: 563–571. (In Russian with English summary).

—— 1979. Simulation of tree growth as an element of forest biogeocoenoses. *Problems of Cybernetics* 52: 73–111. (In Russian).

Rastetter, E.B. and Shaver, G.R. 1992. A model of multiple-element limitations for acclimating vegetation. *Ecology* 73: 1157–1174.

Razumovsky, S.M. 1981. *The Laws of Biocoenoses' Dynamics*. Nauka. Moscow. 232 p. (In Russian).

Reed, K.L. 1980. An ecological approach to modeling growth of forest trees. *Forest Science* 26: 33–50.

Rhynsburger, D. 1973. Analytic deliniation of Thiessen polygons. *Geographical Analysis* 5: 133–144.

Ripley, B.D. 1976. The second-order analysis of stationary processes. *Journal of Applied Probability* 13: 255–266.

—— 1979. Tests of 'randomness' for spatial point pattern. *Journal of Royal Statistical Society* B 41: 368–374.

Ross, Ju.K. 1975. *Radiation Conditions and Architectonics of Vegetation Cover*. Hydrometeoizdat. Leningrad. 342 p. (In Russian).

Rubzov, V.I., Novoselzeva, A.I., Popov, V.K. and Rubzov, V.V. 1976. *Biological productivity of Scots pine in forest-steppe zone*. Nauka. Moscow. (In Russian).

Runge, E.C.A. 1973. Soil development sequences and energy models. *Soil Science* 115: 183–193.

Ryzhova, I.M. 1993a. The analysis of stability and bifurcation of carbon turnover in soil-vegetation systems on the basis of the nonlinear model. *Systems Analysis Modelling Simulation* 12: 139–145.

—— 1993b. Analysis of sensitivity of soil-vegetation systems to variations in carbon turnover parameters based on a mathematical model. *Eurasian Soil Science* 25: 43–50.

Semevsky, F.M. and Semenov, S.M. 1982. *Mathematical Modelling of Ecological Processes.* Hydrometeoizdat. Leningrad. 280 p. (In Russian).

Sennov, S.N. 1984. *Ecological Basis of Tending Forests.* Forest Industry Publ. Moscow. (In Russian).

Serebriakova, T.I. (ed.). 1977. *Coenopopulations of Plants. Development and Interrelations.* Nauka. Moscow. 134 p. (In Russian).

Sharp, F.R. and Lotka, A.J., 1911. A problem in age distribution. *Philosophical Magazine* 21: 435–438.

Sharpe, P.J.H., Walker, J., Penridge, L.K. and Wu, H. 1985. A physiologically-based continuous-time Markov approach to plant growth modelling in semi-arid woodlands. *Ecological Modelling* 29: 189–213.

Shugart, H.H. 1984. *Theory of Forest Dynamics.* Springer. Berlin. 278 p.

Shugart, H.H., Leemans, R., and Bonan, G.B. (eds.). 1992a. *A System Analysis of the Global Boreal Forest.* Cambridge University Press. Cambridge.

Shugart, H.H., Smith, T.M. and Post, W.M. 1992b. The potential for application of individual-based simulation models for assessing the effects of global changes. *Annales Review Ecological Systems* 23: 15–38.

Shugart, H.H. and West, D.C. 1977. Development of an Appalachian deciduous forest succession model and its application to assessment of the impact of the chestnut blight. *Journal of Environmental Management* 5: 161–179.

Shugart, H.H. and West, D.C. 1981. Long-term dynamics of forest ecosystems. *American Scientist* 69: 647–652.

Shugart, H.H., West, D.C. and Emmanuel, W.R. 1981. *Patterns and dynamics of forests: An application of simulation models.* In: Shugart, H.H., West, D.C. and Botkin D.B. (eds.). Forest Succession: Concept and Application. Springer-Verlag. New-York. Pp. 74–94.

Siitonen, M. and Nuutinen, T. 1996. *Timber production analyses in Finland and MELA system.* In: Päivinen, R., Roihuvuo, L. and Siitonen, M., (eds.) Large-Scale Forestry Scenario Models: Experiences and Requirements. EFI Proceedings No. 5. European Forest Institute. Joensuu, Finland. Pp. 89–98.

Silander, J.A. Jr. and Pacala, S.W. 1985. Neighborhood predictors of plant performance. *Oecologia* 66: 256–263.

Silander, J.A. Jr. and Pacala, S.W. 1990. *The application of plant population dynamic models to understanding plant competition.* In: Grace, J.B. and Tilman, D. (eds.). Perspectives on Competition. Academic Press Inc. San Diego, California. Pp. 67–91.

Smeck, N.E., Runge, E.C.A. and Mackintosh, E.E. 1983. *Dynamics and genetic modelling of soil systems.* In: Pedogenesis and Soil Taxonomy. 1. Concept and Interaction. Amsterdam. Pp. 51–81.

Smirnova, O.V., Chistiakova, A.A., Popadyuk, R.V. et al. 1990. *Population Organization of Vegetation in Forest Regions.* Academy of Sciences of the USSR. Pushchino. (In Russian).

Smirnova, O.V., Chistyakova, A.A., Zaugolnova, L.B., Evstigneev, O.I. and Popadiouk, R.V. 1998. Ontogenetic concept of the tree structure and functioning. *Trees* (in press).

Smith, O.L. 1979. An analytical model of the decomposition of soil organic matter. *Soil Biology and Biochemistry* 11: 585–606.

Smith, P., Smith, J. and Powlson, D.S., (eds.) 1996. *GSTE Task 3.3.1 Soil Organic Matter Network (SOMNET): 1996 Model and Experimental Data.* GSTE Report No 7. Wallingford, UK. 255 p.

Smith, P., Smith, J.U., Powlson, D.S., McGill, W.B., Arah, J.R.M., Chertov, O.G., Coleman, K., Franko, U., Frolking, S., Jenkinson, D.S., Jensen, L.S., Kelly, R.H., Klein-Gunnewiek, H., Komarov, A.S., Li, C., Molina, J.A.E., Mueller, T., Parton, W.J., Thornley, J.H.M., Whitmore, A.P. 1997. A comparison of the performance of nine soil organic matter models using datasets from seven long-term experiments. *Geoderma* 81: 153–225.

Smith, T.M. (ed.). 1996. The Application of Patch Models of Vegetation Dynamics to Global Change Issues. *Climatic Change (Special Issue) 34.* 314 p.

Smith, T.M. and Urban, D.L. 1988. Scale and resolution of forest structural pattern *Vegetation* 4: 143–145.
Solbrig, O.T. 1981. Studies in the population biology of genus *Viola*. II. The effect of plant size on fitness in *Viola sororia*. *Evolution* 35: 1080–1093.
Solomon, A. and Shugart, H.H. 1984. *Integrating forest-stand simulations with pàleontological records to examine long-term forest dynamics.* In: Ågren, G.I. (ed.), State and Change of Forest Ecosystems. Uppsala. Sweden. Pp. 333–357.
Sterner, R.W., Ribic, C.A. and Schatz, G.E. 1986. Testing for life historical changes in spatial patterns of four tropical tree species. *Journal of Ecology* 74: 621–633.
Stoyan, D., Kendall, W.S. and Mecke, J. 1987. *Stochastic Geometry and its Applications.* Springer. Berlin.
Strauss, D. 1975. A model for clustering. *Biometrika* 62: 467–475.
Sukhachev, V.N. 1975. *Problems of Phytocoenology.* Nauka. Leningrad. (In Russian).
Sukhovolski, V.G. 1996. *Modelling of tree growth and inter-connection between forest insects and tree plants: an optimisation approach.* Thesis of Dr. Sci. in Biophysics. Forest Institute of Siberian Branch of Russian Academy of Sciences. Krasnoyarsk.
Sverdrup, H.U. 1990. *The Kinetics of Base Cation Release Due to Chemical Weathering.* Lund University Press. Lund. 246 p.
Sverdrup, H., Warfvingle, P. and Goulding, K. 1995. Modelling recent and historical soil data from the Rothamsted Experimental Station, UK using SAFE. *Agricultural Ecosystems and Environment* 53: 161–177.
Svirezhev, Yu.M. and Logofet, D.O. 1978. *Stability of Biological Communities.* Nauka. Moscow. 252 p. (In Russian, English translation: 1983, Mir, Moscow).
Targulian, V.O. 1986. Some theoretical problems of pedology as an earth science. *Soviet Soil Science, Moscow* 12: 107–116 (In Russian with English summary).
—— 1987. *Soil as a planetar surface cover of biospheric planet.* In: Advances in Earth Science. Nauka. Moscow. Pp. 67–92. (In Russian with English summary).
Terskov, I.A. and Terskova, M.I. 1980. *Growth of Evenaged Stands.* Nauka. Novosibirsk. 206 p. (In Russian).
Thomas, M. 1949. A generalization of Poisson's binomial limits for use in ecology. *Biometrika* 36(1): 18–25.
Thornley, J.H.M. 1976. *Mathematical Models in Plant Physiology.* Academic Press. London.
Thornley, J.H.M. and Cannell, M.G.R. 1992. Nitrogen relations in a forest plantation – soil organic matter ecosystem model. *Annals of Botany* 70: 137–151.
Tiktak, A. and Van Grinsven, H.J.M. 1995. Review of sixteen forest-soil-atmospheric models. *Ecological Modelling.* 83: 35–53.
Tiurin, I.V. 1937. *Soil Organic Matter.* Selkhozgis. Moscow; Leningrad. 288 p. (In Russian).
Tomppo, E. 1986. Models and methods for analyzing spatial pattern of trees. *Communicationes Instutae Forestalia Fenniae* 138: 1–65.
Treskov, S.A. 1987. *On higher plant model.* In: All-Union Workshop on Theoretical Plant Morphology. Lvov. Pp. 80–83. (In Russian).
Tucker, S.L. and Zimmerman, S.O. 1988. A non-linear model of population dynamics containing an arbitrary number of continuous structure variables. SIAM. *Journal of Applied Mathematics:* 3549–3551.
Tuzinkevich, A.V. 1988. Modeling of Plant Population Dynamics on the Base of Integro-Differential Equations. Problems of Ecological Monitoring. *Hydrometeoizdat, Leningrad* 11: 276–285. (In Russian).
Ulrich, B., van der Ploeg, R.R. and Preuzel, J. 1979. Matematische Modelierung der Funktionen des Bodens im Stoffhaushalt von Okosystemen. *Zeitschrift für Pflantzenerhnarung und Bodenkunde* 142: 259–274.
Uranov, A.A. 1935. *About contingency of elements in plant community.* In: Proceedings of the Department of Natural History of Lenin Moscow State Institute of Teachers Training. Issue 1: 59–85. (In Russian).

—— 1955. Quantitative description of interspecies relations in plant community. *Bulletin of Moscow Society of Investigators of Nature Ser. Biol.* 60: 31–48 (In Russian).
—— 1965. *Phytogenic field.* In: Problems of Soviet Botany 1. Proceedings of 3rd All-Union Botanical Congress. Nauka. Moscow; Leningrad. Pp. 251–254. (In Russian).
—— 1975. Age spectrum of phytocoenopopulations as a function of time and energy wave processes. *Biological Sciences, Moscow* 2: 7–34. (In Russian with English summary).
Uranov, A.A. and Mikhailova, N.F. 1974. The experience of investigating of phytogenic field of *Stira rennata* L. *Bulletin of Moscow Society of Investigators of Nature, Ser. Biol.* 79(5): 151–159. (In Russian with English summary).
Urban, D.L. and Smith, T.M., 1989. Extending individual-based forest models to simulate large-scale environmental patterns. *Bulletin of the Ecological Society of America* 70: 284.
Urban, D.L., Bonan, G.B., Smith, T.M. and Shugart, H.H. 1991. Spatial applications of gap models. *Forest Ecology and Management* 52: 95–110.
Van Tongeren, O. and Prentice, C. 1986. A spatial simulation model for vegetation dynamics. *Vegetatio* 65: 163–173.
Van Veen, J.A., Ladd, J.N. and Frisse, M.J. 1984. Modelling C and N turnover through the microbial biomass in soil. *Plant and Soil* 76: 257–274.
Vasilevich, V.I. 1969. *Statistical Methods in Geobotany.* Nauka. Leningrad. 230 p. (In Russian).
Verberne, E.L.J., Hassing, J., de Willigen, P., Groot, J.J.R. and van Veen, J.A. 1990. Modelling organic matter dynamics in different soils. *Netherlands Journal of Agricultural Science* 38: 221–238.
Volterra, V. 1931. *Lecons sur la Theorie Mathematique de la Litte Pour la Vie.* Gautier Villars. Paris. 214 p.
Voronoi, G. 1908. Nouvelles applications des parametres continus a la theorie des formes quadratiques. Deuxieme Memoire Recherches sur les Paralleloedrus Primitivs. *Journal Reine Angew. Mathematik* B.134: 198–287.
Walker, J., Sharpe, P.J.H., Penridge, L.K. and Wu, H. 1989. Ecological field theory: the concept and field test. *Vegetatio* 83: 81–89.
Walse, C., Blank, K., Bredemeier, M., Lamersdorf, N., Warfvinge, P. and Yi-Jun Xu. 1998. Application of the SAFE model to the Solling clean rain roof experiment. *Forest Ecology and Management* 101: 307–317.
Watt, A.S. 1947. Pattern and processes in the plant community. *Journal of Ecology* 35: 1–22.
Webb, G. 1984. A semigroup proof of the Sharpe-Lotka theorem. *Lecture Notes in Mathematics* 1076: 254–268.
—— 1985. *Theory of non-linear age-dependent population dynamics.* Monographs and Text Books in Pure and Applied Mathematics Seria 89. Dekker.
Weiner, J. 1982. A neighborhood model of annual-plant interference. *Ecology* 63: 1237–1241.
—— 1984. Neighborhood interference amongst *Pinus rigida. Journal of Ecology* 72: 183–195.
—— 1990. Asymmetric competition in plant populations. *Trends in Ecology and Evolution* 5: 360–364.
West, D.C., Shugart, H.H. and Botkin, D.B. (eds.). 1981. *Forest Succession – Concepts and Applications.* Springer-Verlag. New York.
White, J. (ed.) 1985. *The Population Structure of Vegetation.* Dordtrecht. Boston, Lancaster. 626 p.
Wilde, S.A. 1958. *Forest Soils.* John Wiley and Sons. New York.
Wit, de C.T. 1960. *On competition.* Russian translation in: Mechanisms of Biological Competition. 1964. Mir. Moscow. Pp. 395–412 (In Russian).
Wixley, R.A.J. 1983. An elliptical zone of influence model for uneven-aged row crops. *Annals of Botany* 51: 77–84.

Wu, H., Sharpe, P.J.H., Walker, J. and Penrige, L.K. 1985. Ecological field theory: a spatial analysis of resource interference among plants. *Ecological Modelling* 29: 215–243.
Wu, H., Malafant, K.W.J., Penridge, L.K., Sharpe, P.J.H. and Walker, J. 1987. Simulation of two-dimensional point patterns: application of a lattice framework approach. *Ecological Modelling* 38: 299–308.
Wu, J. and Levin, S.A. 1997. A patch-based spatial modeling approach: conceptual framework and simulation scheme. *Ecological Modelling* 101: 325–346.
Warfvingle, P. and Sverdrup, H. 1992. *Hydrochemical modelling.* In: Warfvingle, P. and Sanden, P., (eds.). Modelling Acidification of Groundwaters. SMHI. Norrköping.
Yaalon, D.H. 1975. Conceptual models in pedogenesis: can soil-forming functions be solved? *Geoderma* 14: 189–205.
Zagreev, V.V. et al. 1992. *Reference Book on All-Union Forest Growth Tables.* Kolos. Moscow. 495 p. (In Russian).
Zaslavsky, B.G. and Poluektov, R.A. 1988. *Management of Ecological Systems.* Nauka. Moscow. 295 p. (In Russian).
Zaugolnova, L.B., Zhukova, L.A., Komarov, A.S. and Smirnova, O.V. 1988. *Coenopopulations of Plants.* Essays of Population Biology. Nauka. Moscow. 183 p. (In Russian).
Zech, W., Senesi, N., Guggenberger, G., Kaiser, K., Lehmann, J., Miano, T.M., Miltner, A. and Schroth, G. 1997. Factors controlling humification and mineralization of soil organic matter in tropics. *Geoderma* 79: 117–161.
Zeide, B. and Pfeifer, P. 1991. A method for estimation of fractal dimension of tree crowns. *Forest Science* 37: 1253–1265.
Zheng, D.W., Bengtsson, J. and Ågren, G.I. 1997. Soil food webs and ecosystem processes: Decomposition in donor-control and Lotka-Volterra systems. *American Naturalist* 149: 125–148.
Zhukova, L.A. and Komarov, A.S. 1990. Plurality of ontogeny ways and the dynamics of plant coenopopulations. *Journal of General Biology* 51 (4): 450–461. (In Russian).
Zhukova, L.A. and Komarov, A.S. 1991. Quantitative analysis of the dynamics of the plurality of ontogeny ways in coenopopulations of plantain under different population densities. *Scientific Reports of Higher Education, Biological Sciences, Moscow* 8: 51–67. (In Russian).
Ziman, J.M. 1960. *The Theory of Transport Phenomena in Solids.* Clarendon Press. Oxford.
Zipf, G.K. 1949. *Human Behavior and the Principle of Least Effort.* Addison-Vesley. Cambridge, MA.

Printed in the United States
By Bookmasters

Printed in the United States
By Bookmasters